76th 2014 B

Bill Perkins

Gift from Dan's family

JOHN DEERE

YESTERDAY & TODAY ™

ROBERT N. PRIPPS

DOUG MITCHEL • MARCI McGRATH • CHRIS SMITH

FOREWORD BY ORION SAMUELSON

WEST SIDE
PUBLISHING

Author **Robert N. Pripps** is widely regarded as one of the world's leading tractor historians. Born on a northern Wisconsin farm, the young Pripps became fascinated with farm machinery at an early age (losing his right thumb in the process). He became a licensed pilot and later worked as an aircraft engineer, retiring at 55 in the 1980s. Since then, he has authored more than a dozen books on tractors. Currently, he and his wife Janice live near the original Pripps Wisconsin homestead, where every spring he and one of his three sons produce about 150 gallons of maple syrup.

Foreword writer **Orion Samuelson** has been a fixture on Chicago radio for over 50 years and is America's best-known agricultural broadcaster. After radio stints in Appleton and Green Bay, Wisconsin, Samuelson joined WGN in 1960, where he has hosted a variety of radio and television agribusiness shows ever since. Inducted into the National Radio Hall of Fame in 2003, he currently can be heard alongside Max Armstrong on *The Morning Show* and seen on *This Week in Agribusiness.*

Contributing Writers

Historian and researcher **Marci McGrath** holds a master's degree in history from the University of New Orleans. She has contributed to more than 20 nonfiction books, and is an accomplished musician.

Writer/photographer **Doug Mitchel** has authored over two dozen books, among them *Anatomy of the John Deere.* His photography has graced books, calendars, and the covers of countless periodicals, including *Collectible Automobile®, Old Bike Journal,* and *Route 66.* Doug lives outside Chicago.

Chris Smith is a New Orleans-based writer and editor, and author of "Three Things," an award-winning column on editing and style. His work on history, technology, business, travel, and education has appeared in numerous books, magazines, and other publications.

Factual Verification by Kathryn L. Holcomb.

Special thanks to Max Armstrong, Steve Ballard, Loreesa Barton, David Bordner, Neil Dahlstrom, Robert Elzey, Dwight Hetletved, Mark Igleski, Brian Kunzog, Mike Kroenke, Glen A. Martin/Martin's Bike Shop, Inc., Vivian Merkel, Andrew Morland, Craig Pugsley, Andrew Raimist, Balaji Rengarajan, Ben Rogers, Gloria Samuelson, Paul Schmidt, and Art Weade.

Cover painting, *New Era,* by Charles Freitag.

ISBN-13: 978-1-60553-486-2
ISBN-10: 1-60553-486-2

Manufactured in China.

8 7 6 5 4 3 2 1

Library of Congress Control Number: 200993888

The Model D of 1923, the machine that established Deere & Company as a major player in the tractor business.

CONTENTS

John Deere

Deere promotional book

Model B

Model 4440

Commemorative Barbie

Model 9030

Change, Legacy, and a Dream

Orion Samuelson

The headline on the full-page ad in the farm magazine reads, "One man cultivates two rows at a time; one man plows 5 to 8 acres in 10 hours. It is the answer to the farm labor problem."

The Moline Plow Company of East Moline, Illinois, placed that ad, and in 2003 a farm listener in eastern Iowa sent it to me. According to the copy, the plow (which competed with products from John Deere) was a spectacular improvement in farm mechanization that would revolutionize the world of agriculture. The date of the magazine was April 1923, not quite 90 years ago.

It is mind-boggling for me to look at how dramatically the farm equipment industry has developed in a relatively short time. As you will learn in this book, the company founded by John Deere in the 1830s, after the success of his steel-tipped plow, has played a leading role in the transition of agriculture from back-breaking hand labor to the living-room comfort of an air-conditioned tractor or combine cab with all the latest electronic gear that enables a farmer to trade products on the world market, and use satellite global positioning to make sure he, or she, is indeed plowing a straight furrow.

Equally amazing in this day of mergers and acquisitions is the fact that Deere & Company has continued under the same name for more than 170 years!

In my 75 years, I have lived much of this change, including my days on a Wisconsin dairy farm that didn't get "hooked up" to the rural electric line until 1948. I remember the date, April 11, very well because that day brought a refrigerator and an electric iron into the house, as well as a milking machine into the barn—to say nothing about the bulbs that replaced the kerosene lamps and lanterns, and made study and homework so much easier.

My grandchildren can't begin to comprehend life without electricity, or farming without tractors and combines. That's why this book is so important. It will help my grandchildren and generations to come understand and appreciate the imagination, creativity, and hard work needed to develop and refine the machines that make America's farmers the most efficient in the world.

That efficiency allows 98 percent of us in this country to pursue our careers and dreams because the other 2 percent are the farmers and ranchers who make sure we don't have to be concerned about searching for or producing food.

I'm sure John Deere was a dreamer, but I'm also sure he couldn't have imagined the legacy he and his plow would leave for us and for the whole world. This book captures that legacy, so read and learn, and remember... if you eat, you're involved in agriculture!

Orion Samuelson

—*Orion Samuelson*

Orion Samuelson (*right*) with American Farm Bureau president Charles Shuman in 1964.

Samuelson (*right*) interviews an official of the Wisconsin State Fair.

For generations of farming Americans, a certain green and yellow tractor has been a familiar part of daily life. Some children have grown up saying "John Deere" more often than their uncle's name. In the early 1900s, at the same time that Henry Ford's automobiles were transforming city life, a new generation of Deere machines—tractors fueled by gasoline—were powering a move by countless rural families away from horse-powered farming. The greater prosperity that accompanied this switch to user-repairable, versatile John Deere tractors also helped to build the brand loyalty that the company's marketers reinforced through community events and Deere Day films.

Introduction

Once Upon a Time in America

America the beautiful: Blessed from sea to shining sea. For more than 170 years, equipment from the company founded by plow-maker John Deere has not only been at work on the "fruited plains," but also in the majestic "purple mountains." In fact, from the rice farms of Louisiana to the wheat fields of Nebraska to the orchards of the Northwest, and all around the world, John Deere's tractors, harvesters, 'dozers, and lawn mowers are very much in evidence, as are indeed, his plows!

John Deere, himself, migrated from Vermont to Illinois to set up his blacksmith shop in 1836. At that time, bison still roamed the Great Plains and Indians were still battling the U.S. Cavalry. Women wore high-laced boots and the more civilized towns hosted Sunday evening band concerts. It was in 1837 that John Deere struck upon the idea of using a broken steel saw blade to make a plow; the first of its kind that worked well in the rich, heavy "gumbo" soil of northern Illinois. It was this plow

that launched John Deere's company. This book, then, is the story of nearly 200 years of a great American institution—Deere & Company, which tentatively entered the tractor business in 1912, and jumped in wholeheartedly in 1918. Since then, the name "John Deere" has almost become interchangeable with the word "tractor." Like Kleenex and Frigidaire, the name has instant recognition the world over.

The Deere name, however, has also graced a complete line of agricultural and industrial equipment, as well as a large variety of lawn and garden tractors and tools designed for homeowners.

Before the time of the tractor, America ran on an animal-centered economy that was powered by horses and oxen. In 1918, as the Deere Board of Directors reconciled the company's decision to pin its future on tractors, Willard Velie (pronounced *Veely*) a board member (and John Deere's grandson) was quoted in meeting minutes as saying, "I think it's safe to eliminate the horse, the mule, the

Author and Deere historian Robert N. Pripps, aboard his 1948 Model B.

bull team, and the woman, so far as generally furnishing motive power is concerned."

This was probably as politically incorrect then as now, but definitely made the point that Deere should devote itself to tractor development.

From the beginning of time to 1830, when Cyrus McCormick invented his famous wheat reaper, the human and animal labor that went into producing a loaf of bread was mind-boggling. First, the soil had to be tilled by hand, or with an animal-powered plow (which also had to be fashioned by hand). Next, the seed was sown by hand. When the wheat grew and ripened, cutting was done with a sickle or scythe. Then, the cut wheat was threshed with a flail, or by sharp-footed animals that were led over it as it lay on the threshing floor. It then had to be winnowed by hand to separate the wheat from the chaff. Finally, the grain had to be milled into flour for baking. This, too, was done by hand, until the invention of animal and water-powered mills. The bread dough was kneaded by hand, and finally baked by fire from fuel gathered and processed by hand. That John Deere's plow also came along at

that time, facilitating agriculture on the prairies, was more than fortuitous. It was a revelation, a game-changer. And the plow was followed by the inventions of a myriad of tools and equipment for the farmer, and finally, by the tractor.

It is remarkable that today's devotion to John Deere products, especially by farmers, borders on fanaticism. To get married, or to leave on the honeymoon, on a John Deere tractor is not unusual. Also not unusual is "Tractor Day" at farm-country high schools, when farm boys and girls drive their farm's biggest John Deere to school. Of course other brands are allowed, but are often relegated to the back of the parking lot.

What is it about this company and its products that has propelled it to this envious stature over nearly 180 years? What is the John Deere "mystique" and what has fostered such a loyal following? First, of course, is talented management: Until the 1950s the company was managed by family members (and even today, family members are important stockholders). Especially before the days of incorporation, the family's fortune was on the line with every major business decision. Next comes dedicated employ-

ees: It is not uncommon for a Deere employee to have 30 or even 40 years with the company. The man appointed as company president in 2009, Sam Allen, had been with the company for 34 years. Finally, Deere's dealerships: Many have served their customers since the 1920s. These dealers are the face and voice of the company to their customers, and sell as many intangibles—including trust and good will—as they do tractors and related equipment.

This, then, is a company that typifies an American way of life—a slice of Americana. This book, rather than being like a trip on an interstate, is rather more of a meander down the side roads of Deere & Company's history. We are confident it will give the reader an enjoyable journey.

Robert N. Pripps
Springstead, Wisconsin

In this classically American tableau from about 1940, a farmer works his fields with his Deere Model B.

Furnishing a Superior Product at a Low Cost

"Under the spreading chestnut tree
the village smithy stands"

—Henry Wadsworth Longfellow

JOHN DEERE, BLACKSMITH

John Deere was a legend in his own time. The enterprise he founded in 1837 is also a legend. Deere & Company is one of America's oldest companies in continuous business. It has survived political, economic, and business crises from well before the Civil War and into the 21st century. This was accomplished through a series of gifted leaders, outstanding technical talent, loyal employees and dealers, and by a relentless attention to quality.

John Deere was born February 7, 1804, in Rutland, Vermont, and grew up in nearby Middlebury. His father, William Rinold Deere, was a tailor who had come from England. His mother, Sarah Yates Deere, was the daughter of a British soldier in the Revolutionary War who stayed on after the fighting. When John was four years old, his father made

At 17, John Deere began a four-year apprenticeship to Captain Benjamin Lawrence in his Middlebury, Vermont, blacksmith shop. John Deere used the skills he acquired during the apprenticeship to establish himself as a thriving independent blacksmith.

a trip to England. While waiting for his ship to depart, William Deere wrote a poignant letter home, perhaps out of a sense of foreboding.

In this stylized painting, one of a series by artist Walter Haskell Hinton depicting the life and innovations of John Deere, the blacksmith-turned-inventor and his revolutionary steel plow are at the center of a counterclockwise collage of American agricultural history: early pioneers transporting their lives across the prairie in covered wagons, Deere's walk-behind plows and popular riding Gilpin sulky plow, and the game-changing tractors and attachments that transformed the company into an industrial giant.

Early settlers in the American Midwest could scarcely believe the way the fields in Illinois stretched as far as the eye could see. But there was a big problem: The heavy, black "gumbo" soil that stuck to the cast-iron plows of the time, forcing the operator to stop every few yards and clean the bottom of the plow manually. Midwestern farmers needed a plow that could turn that gumbo soil from one end of the row to another without stopping. In 1837, their problem was solved when blacksmith John Deere turned his ingenuity to a broken steel sawmill blade.

"Take good care of your mother," the elder Deere wrote. It was the last word from him the family ever received. Although the ship and his trunk reached England, William did not. Speculation is that he was washed overboard during the voyage. Sarah Deere continued the tailoring business to support the family, but young John was soon earning money himself. His mother insisted that he attend Middlebury College, which he did for a time. But John was bent toward the practical, rather than the theoretical, so he apprenticed himself to Captain Benjamin Lawrence, a blacksmith.

Contract apprenticeship in the 1800s provided that a young man be taught a trade in exchange for labor at the behest of the journeyman shopkeeper. Besides the instruction in the blacksmithing trade, John Deere received $30 for the first year of the four-year contract, with a $5 increase for each of the following years. He was also taught mathematics, reading, and writing and was given room and board. Workmanship was Captain Lawrence's creed, and it soon became John's, as well. The apprenticeship was completed in 1825, and afterward, Deere, now a journeyman blacksmith, went on to be employed either by others, or in shops of his own.

In his own shops, in Leicester Four Corners and in Hancock, Deere began specializing in tool manufacture. His shovels, hoes, and pitchforks were known for their quality. Deere not only made tools to order, but also made them as stock for later sale, thus gaining insight into the factory business.

In 1827, Deere married Demarius Lamb from Granville, Vermont. But life was hard in Vermont, with too many blacksmiths and not enough business. To make matters worse, his shops twice caught fire and burned to the ground. It was also said of Deere that he spent too much time on the quality of his products to make a profit: He apparently polished the tines of his hay forks to such an extent that they "slipped out of the hay like needles." That was a good recommendation for quality, but not likely a profitable one.

In 1836, with four children and a fifth on the way, John left Demarius and his brood in Vermont and went to Illinois. He traveled by canal boat and stagecoach to Grand Detour, a village about 100 miles west of Chicago. In the period before his departure, Deere had done work in Royalton, Vermont, for a man named Amos

Bosworth. Bosworth had extolled the great opportunities to be had for farmers and blacksmiths on the prairies of Northern Illinois, as he had heard from his friend, Leonard Andrus, who had migrated there from Vermont and started a sawmill. It wasn't long before John Deere was on his way to Grand Detour, a village on the banks of the Rock River; the river makes a horseshoe curve there, encouraging fur traders to give the village its name. Andrus was the initial settler in

Grand Detour, and he reasoned that the river's sweeping "U" turn provided an ideal location to harness the power of the water.

THE MOVE TO GRAND DETOUR
When Deere arrived in Grand Detour he was 32 years old. He had brought with him the necessary tools of his trade, and immediately put them to use. He was able to find work at the Andrus sawmill repairing a broken pitman, a connecting

After a decade spent crafting and selling tools at his shop in Grand Detour, John Deere built a plow factory 75 miles away in Moline, Illinois, where the Mississippi River offered the advantages of material transport and water power.

rod that maintains a saw blade's circular motion.

Soon Deere had his blacksmith shop built and had a good business going. He also built a small house and sent for his family. Demarius arrived carrying their new son, Charles, born after John had departed. Charles would grow up to play an important part in the John Deere Company.

A PLOW THAT SCOURS

Southern Illinois was well populated in the 1700s, but the immense prairies to the north were mostly ignored. The sticky "gumbo supreme" humus soil was incredibly rich in nutrients, but virtually impossible to till because of the thick tangle of grass and roots that eventually moldered down to become more gumbo. This rich, coal-black humus extended downward for more than seven feet in some areas. The thick grass bed choked out most everything else that tried to grow, which left the prairies virtually treeless.

Fortunately, the land was also flat and rock free. The farmers saw the potential of the prairie land, but the task of tilling under the grass bed seemed formidable. Eventually, huge iron plows were developed, pulled by as many as 16 oxen to accomplish what was known as "breaking the prairie." These "breaker" plows were made of cast iron, which is by nature rough and full of surface imperfections known as blowholes. It does not take a polish and is prone to rusting, which further pits the surface.

Once the initial prairie breaking was accomplished, it was necessary for the farmer to re-plow every year to cover the crop trash left after the harvest. This was done with a team of horses and a smaller plow, also iron, which had to be scraped of the gumbo soil every 20 feet or so, since unlike other soils which tended to polish, or scour the plow, the gumbo just stuck. Hence the name "gumbo": It gummed up the works!

John Deere's steel plow changed the scale of American farming in the 19th century as dramatically as automobiles changed cities in the 20th century. Before he invented the steel plow in 1837, vast tracts of fertile but sticky land were left untilled by farmers. Iron plows couldn't adequately handle the soil, and there even was a faction claiming that plowing "poisoned" the earth. That was an extreme position, but there was no arguing about the fact that existing plows often broke in the heavy prairie soil. After the Deere plow of 1837, numberless acres of land could be put to systematic agricultural uses that helped feed the nation.

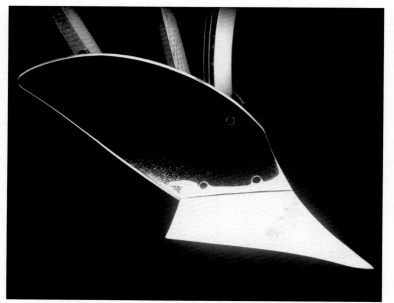

In 1837, after Deere had set up his shop, he spotted a shiny broken steel saw blade lying on the floor of the Andrus sawmill. With a flash of intuition, he saw the shining surface as a solution to the problem of scouring through the gumbo soil. After asking Andrus for the blade, or perhaps a steel share (accounts differ) Deere took it to his shop and fashioned it into a plow. The plow was tried on a neighbor's farm, and it did indeed scour. The integrity of the construction of this plow was dramatically illustrated years later, in about 1900, when a farmer near Grand Detour presented one of Deere's first three plows to Charles Deere. The rugged implement had been in this farmer's family since 1838. The plow was on display in Deere headquarters until 1938, when it was presented to the Smithsonian Institution, where it resides to this day.

After 1837, Deere's business turned more toward the steel plow trade, and away from ordinary blacksmithing. His first plows weren't the huge breaking variety, but rather small plows with about a 12-inch cut. It was the type of plow a farmer would use every spring or fall on land that had already been broken. This small re-tilling plow could be pulled by one or two horses and was light enough to be carried to the field on the farmer's shoulder. As time went on Deere developed new plow shapes and used combinations of cast iron, wrought iron, and steel. To help offset rising financial problems, Deere took in partners, including his friend Leonard Andrus. Henceforth, Deere would leave general blacksmithing behind.

DEERE AND THE GOLD RUSH OF 1849

John Deere moved his operations to Moline, Illinois, in 1848, the year of the stampede for gold that had been discovered in California. This exciting national development affected the young company in several ways—some affecting the company for the good, and others presenting challenges.

On the upside, the new gold strengthened the national banking system. Concurrent was a flow of monies from abroad as opportunities in America blossomed. Further, the western movement of people and goods spurred railroad development. This allowed Deere to expand its marketing territory.

On the downside, men from every walk of life, especially factory workers, abandoned their jobs to head for the "diggings." Thousands came by wagon train across the great prairies; others took the dangerous passage around Cape Horn, while still others crossed the disease-ridden Isthmus of Panama to battle for places on the rickety steamships headed north.

In 1849 alone, the population of California grew from 6,000 to more than 85,000 souls. Most of these did not strike gold, but stayed on anyway, finding the areas around San Francisco and San Jose more hospitable to farming than the windswept prairies. For Deere, this was opportunity, raw but potentially enormous.

A carefully researched replica of young John Deere's blacksmith shop stands today at the John Deere Historic Site at Grand Detour, Illinois. Replicas of Deere's important early plows are displayed outside.

THE NATURE OF AGRICULTURE, 1837–1857

The dawn of the 19th century ushered in the period now known as the Industrial Revolution, which was fueled by several interrelated inventions: the telegraph, the cotton gin, the sewing machine, the Bessemer steel-making process, the McCormick reaper, the threshing machine, and the steam engine. Separately, these would not have amounted to much. But together, they synergistically facilitated each other's development. In agriculture, invention begat invention. The reaper could not come to fruition without the threshing machine, the threshing machine without the steam engine, and so forth. International Harvester Company grew mainly from Cyrus McCormick's and William Deering's companies, both of which developed from efforts to perfect new grain-harvesting implements. John Deere's company evolved alongside the others, but with an initial focus on soil tillage. These pioneer inventors and businessmen, by the sheer force of their determination, freed people from the drudgery of backbreaking toil.

Over the history of American agriculture, tillage practices have largely been the main factor in determining the productivity of land. This is illustrated by what may be an apocryphal story of two brother farmers. Brother Number One obtained some nice bottomland acreage, while Brother Number Two had to settle for land that was hilly, rocky clay. Yet, year after year, Brother Two got better crop yields than did Brother One.

"How do you do that, year after year?" asked Number One. "My ground is obviously better than yours."

"That's easy," said Number Two. "You expect your land to work for you; I work my land."

By 1851, Cyrus McCormick's new company sold more than 1,000 reapers in one year, as farmers warmed to the idea of working larger fields.

The mechanical reaper, invented by Cyrus McCormick in the 1830s, promised a dramatic increase on the harvesting end of the crop cycle that John Deere's plows would bring to the front end. The McCormick Reaper enabled a farmer to cut six times as much grain as he could by hand, and was a natural fit for the wide-open spaces of the Midwest.

BUILDING A BUSINESS

An early John Deere publication on general farm mechanics declares that the plow is the most important implement used in seedbed preparation. Its purpose is to pulverize or break up the soil and to admit air and light, two essentials to normal plant growth. The plow covers surface trash or manure, and mixes it with soil, where it will decay and furnish plant food.

The main parts of the plow are the moldboard (the upper part that rolls the furrow over), the share (the pointed bottom portion, generally removable for sharpening, or replacing), the shin (a replaceable leading edge on some moldboards), the landside (the side opposite the share—it runs along the furrow edge to keep the plow generally going straight), and the coulter (the disk that runs ahead of the moldboard, above the point of the share that cuts the sod and trash so that a clean furrow is rolled over).

Although records of the period are sketchy, it seems that John Deere made three more plows from that broken, discarded saw blade in 1837. Local farmers recognized the utility of Deere's plow, but he faced problems obtaining the steel and financing he needed for expansion. Deere sold few plows in 1838, but it was necessary to take on Leonard Andrus as a partner. The firm was called L. Andrus and Company, but John Deere's name was on each plow. A new factory was built and about ten people were employed for the following year, including Samuel Peek, John's nephew, who did selling for the firm. The market for plows was expanding exponentially, and it soon became necessary to take on a series of partners: Each time, the arrangement was reflected in the name of the company, although John Deere's name was still prominent.

It was at this period in American history that rapid industrialization began to take place. Railroads began to proliferate, brought on in part by the growth in the production of pig iron and its refinement into rolled and wrought iron for rails and steam boilers, and into steel for structures. The railroads facilitated the delivery of coal at reasonable prices, and thus, mining began to expand on a large scale in Pennsylvania and Illinois. A third key development was the start-up of the American petroleum industry, a business that would have far-reaching consequences for Deere.

John Deere and his partners faced many challenges during those early days in Grand Detour, not the least of which was the lack of banks or other means to finance the factory expansion and facilitate payment for plows purchased by farmers. It was also becoming painfully obvious to Deere that Grand Detour no longer was an ideal location for his business. The Rock River, while a good source of waterpower, was not navigable by barges and steamboats except in times of high water. Even coal for the forges had to be trucked in by horse-drawn wagons. Similarly, finished products had to be carried out by traveling salesmen using horse-drawn wagons. John Deere, therefore, began searching for a new location for his factory.

Fewer than 30 years after John Deere had left home at 17 to work as a live-in apprentice for blacksmith Benjamin Lawrence, long days for machinists (*above*) and other workers were still common, even for children. The 10-hour workday maximum was mandated in only a handful of states by 1850, and young children often worked in commercial businesses. That year, printers formed the first national union, a harbinger of a stirring labor movement. Although Deere's workforce numbered less than two dozen at the time, the drive to produce thousands of products—compared with dozens in earlier years—would inevitably change the relationship between management and labor.

THE MOVE TO MOLINE, 1848

Moline, Illinois, lies on the south bank of the Mississippi River where the Mississippi runs generally east and west. It is now part of the Quad-Cities metropolitan area comprised of Bettendorf and Davenport, Iowa, and Rock Island in Illinois. In 1848, when John Deere decided to move to Moline, Bettendorf had not yet been established, and Moline had existed for only five years (1848 is the official founding date as the records of 1843 were lost in a fire). The Rock River enters the Mississippi at Rock Island, about 40 miles downstream from Grand Detour. Because the Mississippi was dammed there, mills and factories sprang up along the Illinois bank. The name "Moline" is a derivation of the French word, "Moulin," the "city of mills."

At the time of the move, Deere and Andrus dissolved their partnership; Andrus was content to stay in Grand Detour to operate the Grand Detour Plow Company. Robert N. Tate, who had been in the Grand Detour group, remained with Deere, and John Gould was added. Tate, an English gentleman, was a good factory man, while Gould excelled in financial matters. By September of that year, Tate had the new factory in Moline built and the first ten plows ready for shipment.

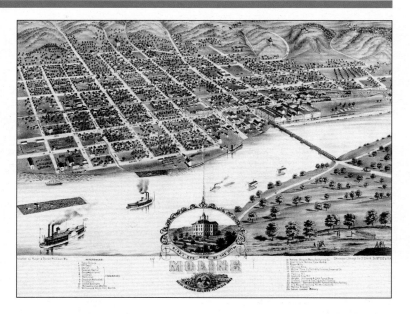

The iron horse and the steel plow: By 1856 rail lines connect the Deere factory to Chicago, New York, and—thanks to a new Mississippi River bridge from Rock Island to Davenport, Iowa—to the entire western United States. This amounted to what would become a profitable marriage of the iron horse and the steel plow.

The firm, now called Deere, Tate & Gould, had specific territories spelled out in an agreement with Andrus that would be honored by the Grand Detour Plow Company salesmen. It seems, however, that these arrangements soon fell by the wayside.

In Moline, the Deere company could get steel and other raw materials directly from St. Louis the year around. Coal was available from local mines. Tate ran the factory, Gould took care of purchasing, financing, and bookkeeping, while John Deere oversaw the marketing and managed the traveling salesmen. Again, partners came and went, most of them arriving with needed infusions of cash. The benefits of asset protection provided by incorporation were not yet widely realized. Whenever any of the partners made a commitment, all were bound by it. In those days, where any form of rapid

When John Deere chose to move his plow factory to Moline, Illinois, in 1848, the small town had just been incorporated. By the late 1880s the fledgling mill town had grown far beyond its original riverfront footprint. The city's Deere-driven prosperity led to a number of urban improvements more common in big cities: electric streetlights, a streetcar system, municipal garbage collection, and public libraries and hospitals.

communication was nonexistent, opposing commitments were bound to arise. Friction, waste, and lost opportunities led to difficulties. Nevertheless, plows got built and delivered to farmers. In November, John Deere's family (with six living children from four to 18 years of age) moved to Moline to stay with the Tate family in their modest home.

Sixteen-year-old Charles Deere joined the firm in 1854, and remained with it for 53 years, until his death in 1907. Charles, who later followed his father as chief executive officer, had an inestimable influence on the firm's success. By 1853, the company, now known as simply "John Deere," produced 4,000 plows annually, plus related tillage implements. By 1856, the railroad and telegraph had arrived in the area, and the company's existence was less tenuous.

Samuel Morse's electric telegraph and patterned "language," the Morse code, would cause American businesses in the second half of the 19th century to rethink their assumptions about communication. By the time of the Civil War, Western Union had built a telegraph line stretching from coast to coast, offering a fast means for placing orders that would benefit Deere's future network of branch distributors.

NAPOLEON OF THE PLOW

In 1857, the local newspaper referred to John Deere with patent hyperbole as the "Napoleon Plow Maker." This was perhaps not as over-the-top a description as it might seem at first glance. In fewer than 20 years, John Deere's company had risen from a one-man operation to one with annual sales of $140,000—not a small number for that era. It was one of the largest, if not *the* largest, plow-maker in America, and there were about 420 of them at that time.

John Deere, himself, was more of a promoter than an inventor. He was astute at anticipating the needs of his customers and being ready with new products while capitalizing on his reputation as the inventor of the steel plow. Over the years, he and his various partners had weathered severe financial and personal difficulties while making the transition from hands-on craftsmanship to big business without losing touch with the farmers who were their customers.

When newspaper editor Horace Greeley wrote in 1871, "Go West, young man, go West and grow up with the country," he might easily have added that John Deere had ventured there already and opened a world of possibilities. Born in Vermont and raised in meager circumstances, Deere parlayed an apprenticeship in blacksmithing, and a knack for adapting tools to the task, into a history-changing enterprise. By the time he died in 1886, at age 82, John Deere had built a manufacturing and distribution empire known across the continent. As he posed for this portrait, he could not have imagined that more than a century later, machines emblazoned with his name could be found working fields in scores of countries around the world.

Chapman and Webber

James Chapman married Jeannette Deere, the daughter of John and Demarius, in 1851. Later that same year, their daughter Ellen married Christopher Columbus Webber. These two sons-in-law would soon join the company, bringing much-needed talents in organization and finance. In 1854, Deere and Chapman entered into a partnership that superseded "John Deere." The plow company would now be called "Deere & Chapman." Webber, whose background was in finance, proved to be an invaluable asset a few years later, during the national economic downturn dubbed the Panic of 1857. Webber maneuvered the Deere family fortune into safe places before the collapse.

THE PANIC OF 1857

National unrest based in disagreement between the states over the issue of slavery followed the bitter election of James Buchanan in 1856. This, in turn, was exacerbated by a sharp, but short-lived financial panic in the summer of 1857. Overbuilding of railroads and overextension of bank credits precipitated the crash. Farm commodity prices fell steeply, causing the farmers to refuse to sell. In turn, farmers didn't have cash to pay for equipment already delivered, nor did banks offer to advance them money for their commodities. Manufacturers such as Deere couldn't pay for raw materials they had already bought. In fact, by the time the situation was fully developed, everyone owed everyone else, but there was no cash to flow.

John Deere began to rely on his sons-in-law for the financial expertise needed to avoid personal and corporate bankruptcy. Employees Luke Hemenway, David Bugbee, and son Charles were all made partners in the firm, alongside John Deere. The firm was now called John Deere & Company. The reorganization was accomplished mostly on paper and with the transfer of some real estate, but the real purpose was to separate John Deere's estate from that of the company so far as was possible under a partnership organization.

Ex-partner John Gould had gone into the banking business in Moline. Twenty-one-year-old Charles Deere approached him for a loan to pay the steel companies for materials already used, and to buy more. Charles offered "accounts receivable" as collateral. Perhaps because of his previous affiliation with John Deere, or because the organization set up by Chapman and Webber made the firm's future look better than it was, the loan was granted and the crisis avoided. From then on, management of the company, called Moline Plow Manufactory, was in the hands of Charles Deere. The partnerships with Hemenway and Bugbee were dissolved.

The Panic of 1857, an economic calamity that included the failure of numerous U.S. banks and railroads, severely tested the young Deere company. After the firm survived a cash-flow crunch and a short-term drop in demand, Charles Deere took over company management, a role he continued to play with great success for more than four decades.

THE NATURE OF LABOR

Prior to the end of the Civil War (1861–1865), the American non-agrarian worker, for the most part, owned his own tools and worked either in his own shop or in a shared two- or three-man shop. Those who worked in larger factories or mills, such as John Deere's Moline operation, were dependent on their employer to provide tools and space to do the work. Workers came from the surrounding towns and farms. Young people had three choices: stay on the farm (if such an opportunity existed), go West (possibly to moil for gold), or find work in a factory, where the average workday was ten to 12 hours and pay for unskilled workers was about a dollar a day.

The first labor unions in America were founded in the 1820s, mainly in the older eastern states. In the main, these unions were local rather than regional. In the West and on the frontier, people were far too independent-minded to think of collective bargaining, especially in plants like the plow works. There are vignettes describing John Deere, himself, pitching in on routine factory work, especially after Charles had taken over day-to-day management.

By the time John Deere invented his revolutionary new plow in 1837, the 363-mile Erie Canal system already linked Lake Erie and the Hudson River. In 1840 more than a million bushels of wheat were floated down the canal to New York. Some of that wheat would certainly have been harvested in Illinois fields worked by farmers using John Deere plows.

By 1840 more than 100 steamboats were in service on the Great Lakes. Early side-wheel versions required huge engines, but the development of more efficient, propeller-driven ships freed more space for people and, especially, cargo. For a time, the steamers were able to hold their own against Erie Canal transport. But the railroads were coming, and by the 1860s steamboats began to decline as a major transportation means to the Midwest and beyond.

By 1850, the ready availability of slave labor in the South, and processing improvements such as Eli Whitney's variation of the cotton gin, combined to create a sobering statistic: Four of every five slaves worked in agriculture, the vast majority in cotton fields. The industrialization of Midwestern farms, including the influence of Deere's steel plow, helped such farms avoid the use of slave labor.

By 1850 more than 9,000 miles of railroad track were in service east of the Mississippi. In the remaining years before the 1861 outbreak of the Civil War, many of the smaller railroad companies merged, and land grants by state and federal governments encouraged larger rail systems. The stage was set for significant use of trains for supply purposes during the war, and for an explosion of commercial transportation after 1865.

Plow Maker to the World

Agriculture at mid-19th-century in America, and indeed around the world, was going through dramatic changes. Older farmland could not compete with the newly developing areas in grain production and so were transitioning from tillage to animal husbandry. John Deere's steel plow was greatly responsible for this in America. Now railroads, steamships, and canals made international distribution of goods possible and the costs reasonable. The prosperity of American farms, due to specialization, caused farmers in Europe to experiment with similar tools and strategies.

Long before Deere tractors existed, many American farmers, such as this one in an 1880 photograph, used conventional horsepower and walking plows to prepare their fields. Soon after moving to Illinois, John Deere realized that plowing in the Midwest was slow going. The heavy "gumbo" soil stuck to cast-iron plow blades and required constant cleaning. His innovation of a smooth-sided steel plow, first fashioned from a broken sawmill blade, was a major breakthrough. For nearly a century, until the 1940s, the Deere company built and sold thousands of smooth-sided steel plows, enabling a dramatic increase in farming production.

The Charles Deere Era

"Let us not forget that cultivation of the earth is the most important labor of man."

—Daniel Webster

"OUR FIRST DUTY IS TO THE PLOW TRADE"

When Charles Deere was just 21 years old, he was taken into the plow company as a partner and its chief executive officer. That partnership, which had been formed as part of a gambit to protect John Deere's personal assets during the financial crisis of 1857, was dissolved later in 1858 after the crisis had passed. The years 1858 and 1859, the first two for Charles, were indeed, trying. Historians say that Charles was changed from an easygoing individual to a stern no-nonsense character with sharp business acumen. Fortunately for the historians, Charles kept a small memo, or "jot-book," covering the years 1858–1869. Although entries during those first two years reveal his trauma over the financial dealings, the jot-book never recorded any criticism of his father. Charles's rather poignant statement, *"I will never, from this seventh day of February, Eighteen Hundred And Sixty A.D. put my name to a paper that I do not expect to pay—so help me God,"* comes close, however.

The business was now solely in Charles's hands, although John Deere was never far from the action. Christo-pher Columbus Webber, Charles's brother-in-law, was made a partner, but Webber soon ran into his own financial difficulties and had to liquidate his share. The company used the official name Moline Plow Manufactory, but unofficially it was known as Deere & Com-pany. The unofficial name became official in 1868.

With railroad service available to the Moline area, marketing for Deere & Company became regional and edged toward the national. To accommodate the expanded trade area, Charles Deere

> **"I will never,** from this seventh day of February, Eighteen Hundred And Sixty A.D. **put my name to a paper that I do not expect to pay**—so help me God."
>
> *—Charles Deere*

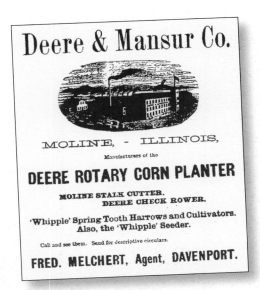

In 1881, Iowa farmers produced more than 81 million bushels of corn, but more was possible with new corn planters just appearing on the scene. The new Deere & Mansur rotary corn planter, introduced in 1879, had a "check-rower," an attachment that used rope knotted at regular intervals to trigger the drop of corn seed. This type of automation was a big improvement on the common farm practice of assigning young boys to hand-drop seed into roughly plowed furrows. County agents such as "Fred. Melchert," who is listed on this 1881 advertisement, played a key role in farming communities by promoting the use of the latest agricultural techniques and tools.

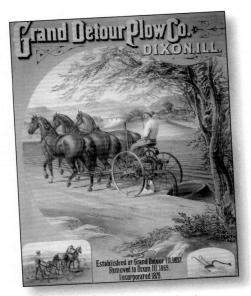

Grand Detour Plow Company, the forerunner to Deere's massive development at Moline, was established near the Rock River in 1837 after John Deere moved from Vermont to Illinois. Deere and his partner Leonard Andrus built a plow factory in Grand Detour when it became too difficult to produce more than 1,000 plows a year from Deere's blacksmith shop. After the partnership broke up and Deere shifted his operation to Moline, the Grand Detour factory later moved to Dixon, Illinois. It became part of the J. I. Case Threshing Machine Company after World War I.

gradually instigated a series of "branch-houses," which allowed the company to take a portion of its traveling salesmen from the field, and install them at the branch houses.

The first branch house was in Kansas City, Missouri, and was jointly owned by Deere and Alvah Mansur. After service in the Civil War, Mansur settled in Colorado Territory. He later helped set up the first St. Louis branch house. A branch house stocked plows of various sizes and configurations, employed salesmen, and provided demonstrations for local farmers.

In the 1860s, Deere & Company began to cautiously expand its product line. Patent rights for the Hawkeye sulky, or riding, cultivator were obtained from its inventor. Examples were built and sent to the branch houses. Eventually, a company catalog was developed to showcase plows, cultivators, planters, and later, buggies and wagons. The branch houses were semi-autonomous and soon began adding equipment from other manufacturers, and even competitors. It was at this point that the stern warning came from Charles Deere: "Don't forget, our first duty is to the plow trade."

Stereoscopic photography—two separate images made to appear three-dimensional to the viewer—was popular in the second half of the 19th century, thanks to innovations such as a handheld model invented by Oliver Wendell Holmes, Sr. This two-panel image of the Moline Plow Works was created around 1875, at a time when the factory was hitting its stride in post-Civil War production with nearly 75,000 plows a year.

A smiling—and riding—farmer operates a Gilpin sulky plow, one of the most successful Deere products in the company's first half century. In this stylized image, company founder John Deere and a trusty dog admire the innovative plow. Note the background depiction at right of a busy Deere factory, presumably turning out more plows, and the appearance of the then-new leaping stag Deere logo imprinted on the plow blade.

BUSINESS AND AGRICULTURE, 1858–1887

As the second half of the 19th century got underway, America was a nation of small farms and businesses, but was becoming a land of big cities and huge business enterprises. Most people still lived on farms, but those places, too, were getting bigger and were marketing to the world. The summer of 1860 saw ample rain and good crops around Moline, and things were looking much better for Deere & Company. European immigrants were finding their way to the

Farmers gather near Winchester, Illinois, for an 1873 Grangers meeting. The Grange movement was founded after the Civil War as an agricultural promotion group. It surged in popularity during the chaos of the Panic of 1873, the start of a six-year economic depression. Grangers used mostly political means to combat what members saw as monopolistic practices of railroads and equipment manufacturers, including Deere. Grangers also started cooperative factories, warehouses, and stores to compete with the industrial giants. As the economy improved in the late 1870s, the Grange movement faded.

As American agriculture became increasingly mechanized after the Civil War, competition among manufacturers intensified. Cyrus McCormick's horse-drawn reaper was a hit, and like John Deere, McCormick built large factories and developed a network of regional agents to sell the new machines. Although some other early Deere competitors faded away or were acquired by Deere, the McCormick threat remained—and intensified in 1902—when financier J. P. Morgan bought McCormick and made it part of his new International Harvester Company.

area and offering themselves both to industry and to the farmers as hired help. Charles Deere felt like he could stop treading water and start wading.

Moline now had ample labor, coal, and steel. Although the Civil War was looming, the strife had not negatively affected the Moline area as seriously as in other regions of the country. Business under Charles was humming. Unlike his father, who took in various partners mostly for the financial infusions they offered, Charles surrounded himself with intelligent, forward-thinking men and, most important, members of the family.

Competition was growing for the plow trade, with more than 2,000 plow-makers by about 1860. One of the most aggressive was the Oliver Chilled Plow Company, founded by James Oliver, a contemporary of John Deere. Oliver, a self-made man in the Deere mold, had amassed $100 by the time he was 32, while doing piecework as a cooper (barrel-maker). While on a visit to South Bend from his home in Misha-waka, Indiana, Oliver was offered one-fourth interest in a foundry at inventory value of $88.96. Since James had his "fortune" in his pocket, he struck the deal and was instantly in the cast-iron plow business.

Oliver, like most rural men of the time, was familiar with plows. He once remarked, "The man who has never been jerked up astride his plow handles, or flung into the furrow by a balky plow has never had his vocabulary tested." He knew that plow technology was ripe for improvement.

The success of Oliver's company came after the 12-year development of a chilling process that didn't warp the casting and left the orientation of the fibers of the metal perpendicular to the surface. This gave the cast iron a wearing resistance and a surface smoothness exceeding that of steel. Soon, J. I. Case and others, including Deere & Company, used chilled cast iron in the manufacture of plows.

As plows and a wealth of other consumer products improved, so too did the outlook for American workers. Indeed, the growth of organized labor in America parallels the expansion of industry. Until after the Civil War, unions were mostly local affairs involving craftsmen (shoemakers, carpenters, etc.), although organization of certain factory "trades" began in the 1850s. The Patrons of Husbandry, or Grangers, involved farmers in a rural social club that helped relieve the isolation of country life, and that agitated successfully for legislative help for farmers, particularly in Iowa, Minnesota, Wisconsin, and Illinois. Like it or not, Deere and other manufacturers were forced to acknowledge that satisfied consumers who felt they were being treated squarely were the keys to long-term company growth.

The Civil War, 1861–1865

Nothing in the history of America before, or since, the Great Civil War has caused the strife and human suffering of that four-year period. The issues of slavery, states rights, and union strife ripped the American nation in half. In the period of fomentation before, and in the aftermath, misery and disruption persisted unabated.

During the War, a Plow City Rifle Company was organized with men from Moline, many of whom had worked for Deere & Company. John Deere's son-in-law, James Chapman, was the only family member to serve in the army during the Civil War, seeing duty as a lieutenant with the Illinois Volunteers until 1864. John Deere himself was pilloried in the more racist papers as a "raging abolitionist," and for calling a debating adversary "a worthless coote...."

Deere & Company was dissolved in 1860, eliminating all partnerships and leaving it in the hands of Charles, alone. He retained the name "Moline Plow Manufactory." In July of 1864, Deere & Company was again reconstituted with Charles and John Deere as equal partners. Despite the national turmoil, business was brisk and profits consistent during the war years. The wartime catalog contained the statement, "We are obliged to hold our plows for cash, or as near as practical, owing to the shortening of credit by iron and steel manufacturers and importers to four months' time." This was an indication that Charles Deere had learned his lesson from the Panic of 1857.

ROCK ISLAND BARRACKS, ILL.

John Deere, the man, had unusual interactions with Rock Island, which sits in the Mississippi about 10 miles northwest of Moline. In 1853 John won two dollars at the First Annual Rock Island Fair for the Best Center Draft Plow. A year later, he chaired the Rock Island County Whig Party convention, and in 1855, with help from a Mr. Thompson, he put out a fire on the Rock Island Bridge. In 1858 Deere and other abolitionists crossed to the island to disrupt a Democratic Party meeting. With the 1861 outbreak of the Civil War, John helped defray the costs of raising a local company of Illinois volunteers. The Union Army used Rock Island as a barracks until 1863, when the first of what eventually swelled to some 12,000 Confederate prisoners arrived on the island. Today, Rock Island is the U.S. Army's largest arsenal.

Incorporation

Prosperity abounded after the Civil War, so it's not known what caused Deere & Company to move to a corporate form of organization, by which the company became a legal "person" that could make contracts in its own right, while limiting the liability of stockholders' personal assets. On August 15, 1868, the firm became an Illinois corporation, retaining the name Deere & Company. Shareholders and officers were John Deere, President; Charles Deere, Vice President; Stephen Velie (John Deere's son-in-law), Secretary; and George Vinton (Demarius Deere's nephew), Treasurer. Charles Deere held 40 percent of the stock, and although listed as vice president, it was his company. In the year after initial incorporation, several others were added as stockholders with small amounts of stock. Among these were Gilpin Moore and C. V. Nason.

Among early Deere employees were three who would play key roles in the business. Gilpin Moore (*front row, second from left*) invented the Gilpin sulky plow, which drove the company's success after the Civil War. Moore held more than 30 patents by the time he retired. In 1863, John Deere made a partner of his son-in-law Stephen Velie (*third from left*), whose financial acumen and focus on product quality helped stabilize the enterprise. Charles Deere (*fourth from left*) was groomed for leadership after his older brother, Francis Albert, died at 18. Charles Deere was an enthusiastic sales leader, often personally demonstrating equipment. He developed Deere's branch distribution network and during four decades transformed his father's plow company into a national concern.

THE TRANSCONTINENTAL RAILROAD

In 1862 President Abraham Lincoln signed a bill authorizing governmental aid to railroad companies to build a "transcontinental" railroad from Omaha, Nebraska, to Sacramento, California. The bill chartered the Union Pacific Railroad to build westward from Omaha, while the Central Pacific would build eastward from Sacramento. Construction was to continue until the tracks met.

Work didn't get started until the end of the Civil War in 1865, but progressed rapidly thereafter. On May 10, 1869, the tracks met at Promontory Point, Utah. Meanwhile, the Chicago & Northwestern Railroad completed a connection from Chicago to Omaha. The first train from California to New York arrived on July 29, 1869, after a run of six and a half days.

The practical effects were quickly felt. Commercial and passenger travel became infinitely easier and more direct, and American trade with the Far East increased 100 percent by 1872. Besides providing Deere & Company with easy access to dealers west of the Mississippi, the transcontinental railroad encouraged John Deere and his wife, Lucenia, to travel regularly to Santa Barbara and San Francisco, where the couple spent their winters beginning in the mid-1870s.

On May 10, 1869, a ceremonial golden spike was driven in Utah to mark the completion of the Transcontinental Railroad. In addition to bringing existing residents of the far West faster, more reliable shipments of farming and mining equipment, the railroad prompted many American farmers to relocate west of the Mississippi to seek cheaper, bigger tracts of land. Once they were established on their new land, they needed plows, planters, and wagons, which Deere was more than happy to provide.

EXPANSION OF THE PRODUCT LINE

The Civil War and its aftermath spurred western population expansion and the extension of the railroad system to the West Coast. Congress passed several "land acts" granting homestead rights to settlers in the various territories, and generous land grants to railroads for acreage that bordered their proposed

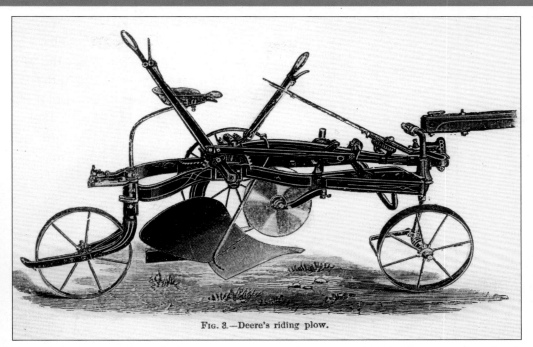

FIG. 3.—Deere's riding plow.

It was only a matter of time before Man's prototypical invention, the wheel, was combined with a Deere plow: This 1880s three-wheeled riding plow let a farmer sit behind horses or, later, a steam tractor, in order to work more land than ever before. The riding plow was one of several innovations that drove a stunning increase in U.S. farming production in the 19th century. By 1900, farmers in California could harvest as much wheat in nine hours as New England farmers had harvested in 150 hours in 1800.

In the year before the Transcontinental Railroad opened in 1869, Deere & Company sold more than 41,000 plows and other items— but that was just a hint of things to come. As western states seized on rail travel to draw settlers to new farmland opportunities, they gave a post–Civil War economic boost to the growing farm implement industry, too.

Charles Deere, the second son born to John and Demarius Deere, led the Deere business through a tumultuous half-century beginning in 1859. Charles, who was sent to a series of good schools, including Bell's Commercial College (Chicago), joined the family firm as a bookkeeper at the age of 16, and quickly showed a knack for finance. Shortly after the Panic of 1857 nearly sank Deere, the founder turned over the helm to Charles, then only 21. Charles Deere would successfully run the company for 49 years, until his death in 1907.

tracks. The Central Pacific route was chartered in 1862, while the Northern Pacific route, the first to be built, was chartered in 1864. The "Golden Spike" linking the transcontinental railroad was driven at Promontory Point, Utah, May 10, 1869. Cross-country rail shipping and travel were possible at last. Farms sprouted on previously virgin land, and every one of those farms needed equipment.

With Deere's development of the branch-house system of distribution, and increased competition in the plow business, Charles Deere was ready to work his company into new areas of endeavor. The Deere catalog of 1866 included 31 implements including various styles of walking and riding plows, walking and riding cultivators, harrows, wagons, buggies, drills, and planters. In time, disc plows and gang (or multiple-bottom) plows were added to the line. Haying and harvesting machinery appeared nearer the turn of the century.

TRANSFORMATION TO AN INDUSTRIAL SOCIETY

One of the great surprises of world history in the late 19th century was the rapid industrialization of the United States of America. The Industrial Age was grounded in the availability of manufacturing power. In the case of the U.S., abundant coal was available for producing power through steam engines and coal-fed plants. This was particularly true after the 1870s, when the vast coalfields of the Appalachians were opened. Later, iron ore deposits were developed in the Mesabi Mountain range of northern Minnesota, Wisconsin, and Michigan.

The huge continental market for manufacturers and farmers that came with rail expansion was helped by—and helped to spur—the postwar resurgence of American banks. Industrial firms were invited to accept credit from banks that once again were well capitalized.

To add to all of this, the labor market was also booming, as immigrants from Europe joined native-born Americans in a significant population shift from farms

Isaac Singer's sewing machine of 1850, one of the first consumer machines to be sold around the world, was emblematic of how mechanization could transform an activity previously performed by hand—and usually by women. On the farm, a similar story was playing out in the fields, as (mostly) men and boys saw their workload eased somewhat by new devices that could prepare the ground, plant the seeds, and harvest the crops.

The Bessemer process, which uses forced air to remove impurities during the production of steel, revolutionized manufacturing in the mid-19th century. The method had been used for centuries in China, but English inventor Henry Bessemer and American inventor William Kelly independently discovered a way to "scale up" the technology and produce large amounts of high-quality steel. For fledgling industrialists like John Deere, the availability of good steel from a U.S.–based plant was a big improvement over rolled steel shipped all the way from England.

to the cities. More city dwellers meant more jobs (or perhaps vice versa), so between 1860 and 1900, the number of wage-earning Americans grew from one million to more than five million.

During the summer of 1871, the Midwest suffered a severe drought that caused a general decrease in farm and farm-machinery profits. Yet Deere & Company didn't suffer. In fact, profits continued to rise. Charles Deere had grown into one of the most respected businessmen in Illinois, and while his company had not seen the explosive growth enjoyed by railroads, he had plotted a steady course of stability and controlled growth that was unaffected by 1871's poor growing season.

Marketing

Under Charles Deere's leadership, marketing of the Deere product line had taken a dramatic turn with the introduction of branch houses. Elaborate catalogs were issued every year and Deere branch houses often developed their own catalogs. Newspaper and magazine ads stressed the "light-draft" of Deere plows. Company stationery from 1857 mentioned additions to the product line, such as "stirring plows, corn plows and ox yokes."

The company also sought publicity by displaying its wares at agricultural fairs. These were mostly local, or county, fairs, but Deere also had a presence at state, national, and international fairs. Such exhibitions garnered magazine publicity. Competitive equipment trials were crowd-pleasing highlights, and prizes awarded were subsequently featured—to profitable effect—in company advertising.

After the Civil War, Deere launched a popular annual publication designed to be given to customers by farm equipment dealers. The 3.5"×6.5" *Farmer's Pocket Companion* usually served as a calendar and notepad, and included farming charts along with, of course, advertisements for the latest John Deere products. The colorful cover of this 1884 edition trumpets an earlier triumph for the company, a grand prize at the Paris Exhibition of 1878 for the Deere riding gang plow.

LABOR RELATIONS

The rise of organized labor unions in the last half of the 19th century affected all of the larger industrial institutions, including Deere. The "strike" was the main tool of the union, and its primary goals were higher wages and better working conditions, including a regular eight-hour day. Bitter strikes, with violence, occurred at mines and railroad-related industries, but Deere & Company enjoyed peaceful labor relations until 1876 when the first strike, by molders, occurred, protesting a reduction in piece-work rates. The strike was "settled" after a week, when the Deere molders came back.

A short time later, Deere announced a 10 percent pay cut for factory workers not engaged in piecework. The workers quickly organized into a union and threatened to strike. Their statement to management was that no violence would occur as long as the company did not receive materials or attempt to ship products until the issue was settled. Deere's competitors seized the advantage offered by this development by announcing to customers that Deere was "shut-down" for the year. Primarily through the local papers, Deere made its case that times were difficult and costs were too high. Eventually, the union accepted the company's terms.

More labor strife occurred in 1884, when Deere grinders went on strike to have their wage cut reversed. This time, the national Knights of Labor was involved, and pressured Deere polishers to support the grinders. Deere was replacing the strikers as fast as it could hire. Shortly thereafter, the infamous Haymarket Riot exploded in Chicago, which, with violence and loss of life, had the effect of discrediting the national unions. A compromise wage settlement was reached a few days after, and peace was restored at Deere & Company.

Charles Deere and other Deere executives, already concerned about the company's growing labor issues, sat up and took notice on May 4, 1886, when a dynamite bomb exploded during a labor protest in the Haymarket District of Chicago. This wasn't just a bomb—it was distant thunder that portended significant change. Union employees had been locked out of the McCormick Reaper Works for three months. Deere had no labor unions at the time, and the company's labor grievances never gained the notoriety of those in Chicago and other large industrial cities of the late 19th century. Nevertheless, the company faced a series of challenges, including a brief, informal strike in 1876. By 1887, competitive pressures forced Deere to begin paying its workers health-and-accident benefits, and the company's traditionally close relationship with workers changed over time as factories grew larger and workers' expectations evolved.

THE END OF AN ERA: THE DEATH OF JOHN DEERE

After incorporation, John Deere's role in the company was minimal. He was more interested in Moline and in politics, it seemed, than in the day-to-day operations of the plow works. He served a term as mayor of Moline, and was on the boards of several banks.

John Deere's wife, Demarius, the mother of his nine children (the tenth was stillborn), died in 1865, just short of her 60th birthday. Later that year, John returned to Vermont to the Lamb homestead to commiserate with the family over her death. While there, he became reacquainted with Demarius's maiden sister, Lucenia. When John Deere returned to Moline, he brought Lucenia with him as his new wife.

John enjoyed his family and revisited familiar surroundings. He and grandson C. C. Webber traveled to Minneapolis in 1885 to enjoy the Minnesota State Fair. After their marriage, John and Lucenia traveled by train to Santa Barbara, California, in the winter months. John sent back glowing reports to Charles about agriculture in the Golden State. Summer months, however, found John and Lucenia back in Moline at their palatial "Red Cliff" mansion.

During the winter of 1885, John Deere's health began to decline, and on May 17, 1886, he died in his sleep at his home in Moline. He was 82 years old. More than 3,000 people (virtually the whole town) attended his funeral in Moline. A floral "plow" graced the coffin with "John Deere" written on the beam.

Although John Deere held some important patents, he was not noted as an inventor. Nor was he a financial genius or a diplomatic leader of people. However, he possessed the charismatic qualities shared by many self-made men that cause people to go out of their way to please. He was a man of great personal dignity and bearing, and he personified the company's dedication to integrity and quality. Such was the nature of the man, and his company, that his name is still displayed on the finest agricultural and industrial equipment available the world over.

The company, by now 50 years old, had progressed from a one-man blacksmith shop that built just three plows its first year, to a multimillion dollar company employing thousands.

Tremendously difficult financial times were weathered, and great strides in organization and factory practices were accomplished. The reins of leadership were seamlessly passed to John's son, Charles, who quickened the tempo of progress begun by his father. John Deere was, and is, an American icon.

John Deere, 1804–1886, Rest In Peace

The "Red Cliff" mansion on 11th Avenue in Moline was John Deere's home for the last few years of his life. The house is shown here in 2008 while awaiting restoration after decades of use as multifamily housing. But from the front porch in his final years, Deere and his second wife, Lucenia, had a panoramic view of the riverfront below and the bustling city that owed its prosperity to his company. Red Cliff is on the National Register of Historic Places.

In 1872, Charles Deere built the Deere-Wiman House, a three-story Victorian mansion on a seven-acre site overlooking the plow factory and the Mississippi River. Today, the John Deere Company operates the home and its gardens for visitor tours and cultural events in Moline.

THE DEERE FAMILY TREE

The colossus that became Deere & Company traces its origins to semi-rural Vermont. Although John Deere's father, William, disappeared while en route to England, when John was just four years old, the boy lived with his mother, Sarah, until he was apprenticed to a blacksmith at 17.

John's willingness to work hard helped ensure his success as a businessman, and as the patriarch of the family that is delineated here.

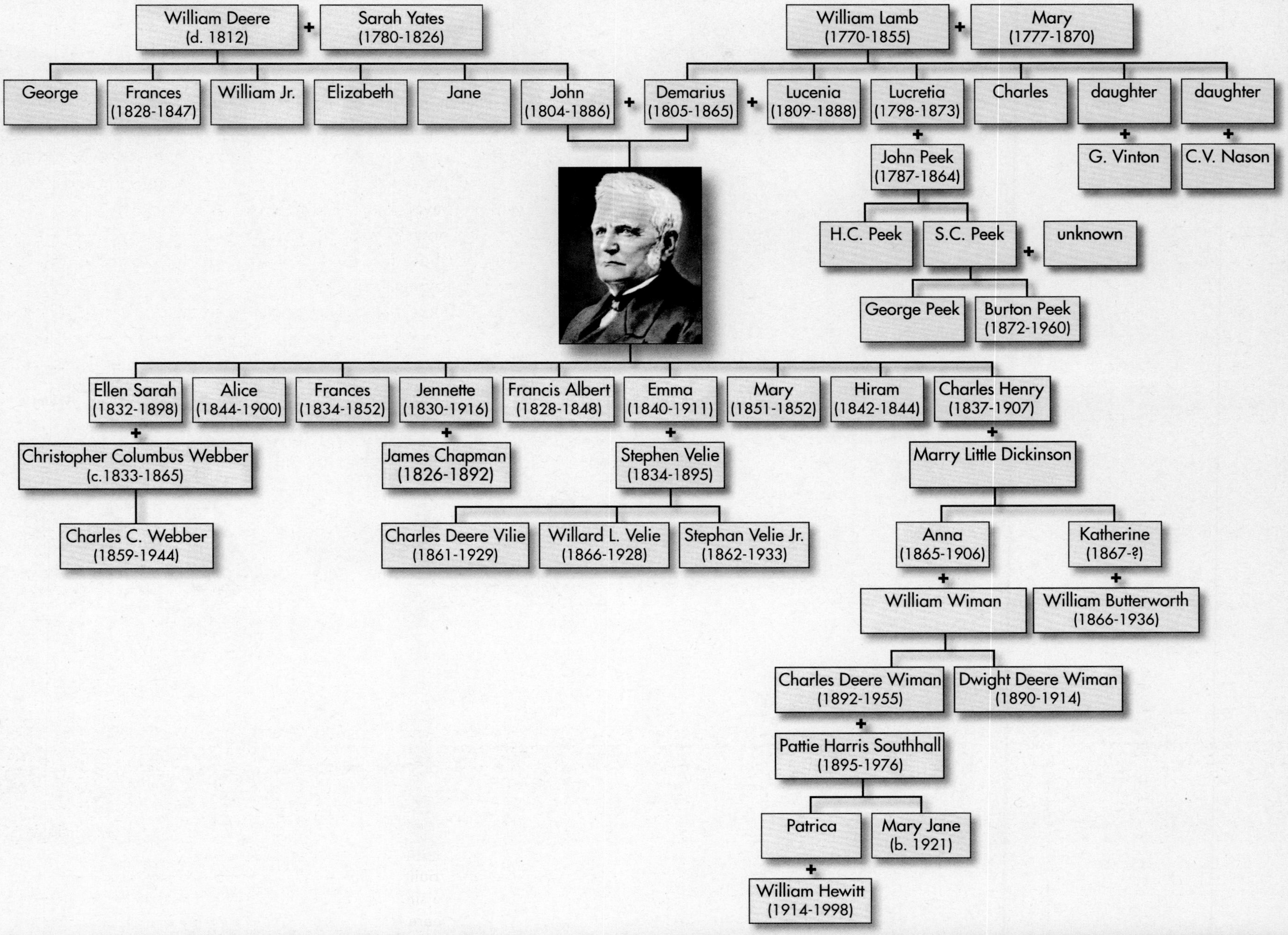

William Deere (d. 1812) + Sarah Yates (1780-1826)

William Lamb (1770-1855) + Mary (1777-1870)

George — Frances (1828-1847) — William Jr. — Elizabeth — Jane — John (1804-1886) + Demarius (1805-1865) + Lucenia (1809-1888) — Lucretia (1798-1873) — Charles — daughter — daughter

John Peek (1787-1864)

G. Vinton — C.V. Nason

H.C. Peek — S.C. Peek + unknown

George Peek — Burton Peek (1872-1960)

Ellen Sarah (1832-1898) — Alice (1844-1900) — Frances (1834-1852) — Jennette (1830-1916) — Francis Albert (1828-1848) — Emma (1840-1911) — Mary (1851-1852) — Hiram (1842-1844) — Charles Henry (1837-1907)

Christopher Columbus Webber (c.1833-1865)

James Chapman (1826-1892)

Stephen Velie (1834-1895)

Marry Little Dickinson

Charles C. Webber (1859-1944)

Charles Deere Vilie (1861-1929) — Willard L. Velie (1866-1928) — Stephan Velie Jr. (1862-1933)

Anna (1865-1906) — Katherine (1867-?)

William Wiman — William Butterworth (1866-1936)

Charles Deere Wiman (1892-1955) — Dwight Deere Wiman (1890-1914)

Pattie Harris Southhall (1895-1976)

Patrica — Mary Jane (b. 1921)

William Hewitt (1914-1998)

The Deere experiment with the three-wheeled, three-plow-bottom Melvin tractor didn't last but helped guide future tractor successes by the company. Worried about falling too far behind in the market, Deere & Company assigned in-house engineer C. H. Melvin to design an affordable tractor.

Melvin's attempt, too-obviously based on a competitor's machine, the Hackney Motor Plow, was abandoned in 1914 as insufficiently strong and reliable. But Melvin's power-lifting device for the plow, and the field tests he supervised from 1912 to 1914, provided valuable lessons.

A Time for Expansion

"I have walked many a weary mile behind a plow, and I know the drudgery of it!"

—*Henry Ford*

MONOPOLIES AND TRUSTS

After the death of John Deere in 1886, the company, and indeed the United States, entered into a period of relative prosperity. This was both good and bad. The general economic growth came hand in hand with an inexorable movement toward corporate bigness that was inspired by intense rivalries in every field of business. Competition forced smaller concerns to merge in order to survive and challenge bigger ones. Robust company size was critical to savings, to the purchase of raw materials, to the ability to acquire patents from inventors, and to dealings with banks and labor. Most important, the drive to increase company size revealed the desire to achieve a monopoly.

During the late 1880s and through the 1890s, the leading entity in a given field might persuade its important competitors to combine into what was then known as a trust: an unchartered "agglomeration" holding the stock of all of the concerns in trust. Those not in the trust were systematically squeezed out of business. A single board of directors managed the typical new mega-corporation. Later, the word "trust" was applied to any company enjoying unfair monopolistic, or near-monopolistic, control of an industry.

In 1889, a syndicate of British investors approached Charles Deere with an offer to buy out Deere stockholders. The syndicate didn't hide the fact that it was simultaneously negotiating with other plow makers. In fact, they made it known that were the deal to go through, they'd work to gain control of Moline Plow and Deere & Mansur. The obvious threat was that if Deere were the only company to resist the takeover bid, it would inevitably face a squeeze from a plow trust. The plow makers already faced pressure from a "Steel Trust," but presented a united front through the Northwestern Plow Association to hold steel prices down and, if that failed, to raise plow prices together. Such trusts, collusion, and price-fixing had not yet been made illegal. After about a year of intrigue and machinations, the British bid for Deere was turned aside. Soon thereafter, in 1890, Congress passed the Sherman Anti-Trust Act.

THE PANIC OF 1893

America was struck by a major financial depression in the summer of 1893. The root cause was the failure of a British bank, Baring Brothers. That collapse forced many English investors to sell

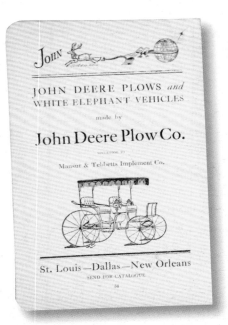

The sale of non-Deere products by Deere's early branch locations sometimes became a testing ground for future company ventures. In the early 1880s, the popularity of "white," meaning unfinished, buggies at the St. Louis branch led Deere to give White Elephant Vehicles nearly equal billing with John Deere Plows in sales catalogs for all of its branches. Deere later purchased the Mansur & Tebbetts wagon factories and incorporated them into the growing Deere business. Wagons would remain an important product for the company well into the 20th century.

their American securities in order to obtain gold. The drain on the U.S. Treasury's store of gold was exacerbated by the Sherman Silver Purchase Act, which required that 4.5 million ounces of silver bullion be purchased by the Treasury each month. Meanwhile, bankers were redeeming their silver certificates for gold. Consequently, the Treasury was running out of gold to back the "greenback" dollar. The national money supply contracted and was followed by massive deflation (a general drop in prices caused by the contraction of available money and credit). Falling prices for goods produced by farmers forced them to cut back their output, leaving Deere and other equipment manufacturers with acres of unsold merchandise. The glut of inventory, in turn, led to an increase in unemployment among factory workers.

Rescue came in 1896, when the U.S. Treasury entered into an agreement with investment banker J. P. Morgan to furnish $62 million in gold in exchange for government bonds. With that, and the repeal of the Silver Purchase Act, confidence in the dollar was restored and business resumed as before.

A turn-of-the-20th-century chuck wagon, stamped "John Deere Wagon Company," is a colorful reminder of the rugged environments in which many of the company's customers worked just a few years before gasoline engines changed everything. Many Deere wagons are highly prized by collectors. This one, which also displays the stenciled name of a previous owner, N. B. Allen & Co., sold for nearly $30,000 at auction in 2007.

The steel lever harrow, designed to be pulled by a team of three or four horses, enabled farmers to pulverize and smooth a field, to trap moisture, and limit weed growth. Deere's 1893 model could be adjusted to provide simple smoothing or penetration of the soil. Before Deere tractors appeared, a harrow was often paired with a riding cart for the driver.

Joseph Dain was on the board at John Deere in 1914 when a majority of directors thought the company should explore the pros and cons of getting into the tractor business. Dain was asked to look into the concept further and report back to the board with his findings. Dain returned in 1915 with a prototype that he thought could be built and sold at a profitable price. Although the board was lukewarm to the idea, Dain was given approval to continue testing. It appeared that his all-wheel-drive tractor could be sold for $1,200, which was an attractive retail price considering the advantages the machine could bring to farmers. After earning board approval to build 100 units of his new tractor, Joseph Dain suddenly died. Before that tragedy, Deere spent $250,000 to build just 10 units. Shortly after, the Dain project was removed from further corporate discussion.

DEERE AND THE LABOR MOVEMENT

The period from 1880 to 1890 saw the rise of militant labor unions across the nation, and activism eventually manifested itself at Deere & Company. In the "old days," John Deere himself was often seen tinkering in the shop, or even helping with the work. Now, the company seemed to be prospering, and Charles Deere and the other top executives lived in luxurious houses overlooking the factory and the homes of the workers. Conditions in the shop were often difficult, especially in the summer. No factory job at Deere was exactly pleasant in those days, but workers in the grinding room had a particularly rough go of it, becoming ill from breathing metal dust all day.

Charles Deere was a staunch Republican who saw the 1892 election of Democrat Grover Cleveland to the presidency as a portent of financial disaster. Cleveland had been president from 1885 to 1889 (he lost the office, temporarily, to Benjamin Harrison in 1888). During his first term, Cleveland established himself as a liberal activist opposed to the unchecked growth of American business. During his second term, Cleveland lowered expectations for sales in the 1890s and systematically reduced prices—and therefore wages. Cleveland's foresight probably saved Deere & Company during the Panic of 1893, but also fomented further labor unrest.

In 1892, Deere hired a Pinkerton Agency detective to work undercover because of the tensions in the grinding department. But because most of the grinders were Swedish, and spoke to each other in their native language, the undercover detective reported that although the workers wanted to strike, they probably lacked the will to do so.

In truth, Deere paid slightly more than the going wages for similar work at other companies. Also, Deere did the best it could, considering the times, to provide good working conditions. Even so, the countrywide labor unrest in mines and railroads spilled over into the Deere factory. The grinders did strike, and were out for several months. When Deere imported both experienced and inexperienced men as replacements, the strikers' bitterness led to physical violence and arrests before Gilpin Moore, the shop superintendent, engineered a settlement.

Samuel Gompers, president of the newly formed American Federation of Labor (AFL), came to Moline in the spring of 1898 urging "truthfulness and cool judgment" in labor disputes. By then, the AFL had organized unions for craft workers, such as barbers and bricklayers, and for factory workers of all types.

BRANCH HOUSES AND TRAVELERS

In the early years of Deere & Company, the marketing of products was centralized in Moline, Illinois. The company sent its own personnel throughout the territory to visit independent retail outlets, such as hardware stores and implement dealers. Deere personnel doing this job were called "travelers." They often lived in the territory they covered and worked from their own homes. This system of marketing was widely used by other agricultural equipment companies as well. Travelers from competing firms got to know each other and were often in direct competition for sales. A limited ability to communicate with the home office forced a great deal of responsibility upon the travelers regarding price and delivery quotations. Head-to-head competition frequently led to unauthorized price cutting and promises of impossible delivery schedules.

To bring in a modicum of local control while retaining the independent spirit of competition, Charles Deere initiated the first "branch house," in Kansas City, Missouri, in 1869. After the Kansas City house proved successful, four more were added, in St. Louis, Minneapolis, Council Bluffs/Omaha, and San Francisco. As originally established, these were independent corporations, with Charles Deere as one of their principals. Branch houses printed their own catalogs and offered products from vendors other than Deere. This sometimes led to bitter competition between branch houses. In one case, a branch catalog offered the Gas Traction Company's Big 4 gasoline tractor as if it were a Deere product. Despite this kind of lapse, the branch-house system worked well enough until ▶

Manufacturing Improvements

At the turn of the century, agricultural manufacturers enjoyed a fecund period of growth and labor peace. Deere & Company had an assembly building that covered nine acres beneath one roof, complete with electric lighting and an overhead-track trolley system that ferried finished items to the warehouse. Cement flooring added to the efficiency of the assembly operations. These improvements, as well as experiments in scientific management techniques, led to dramatic reductions in the cost of assembling plows and loading them into railroad cars.

In 1869 Charles Deere and Alvah Mansur established a "branch" company (*left*) in Kansas City, Missouri, to distribute Deere products in the region. The success of that innovation led to other branch locations in St. Louis, Minneapolis, Council Bluffs/ Omaha, and San Francisco, forerunners of Deere's present-day sales network. The company was one of the first in Kansas City to choose a site for its proximity to rail lines. This building housed storage, offices, and salesrooms.

proposals for a federal income tax began to be floated, and it appeared that the company would one day be penalized with additional taxes due to its decentralization. Branch houses were gradually brought under centralized control, and in 1910 the Deere board of directors passed a resolution that all factories and distribution centers should be unified. This was accomplished by 1911, and the entity that resulted was thereafter referred to, in Deere inner circles, as the "Modern Company." The federal income tax became law in 1913.

The English inventor Jethro Tull revolutionized seed planting in 1701 with his design for a horse-drawn seed drill. His invention led to seed drills like this one, shown outside a John Deere store in 1910 with Nina Hewitt holding the reins. The device, which regulates the depth and spacing of seed planting, helped farmers bring in consistent crops on large tracts of farmland in the American Midwest.

DEERE AND THE COLUMBIAN EXPOSITION

The Paris Exposition of 1889 dazzled the world with its Eiffel Tower and amazing displays of worldwide culture, progress, and opulence. In 1890 the U.S. Congress decided to go the French one better by mounting a World's Fair to celebrate the 400th anniversary of the "Discovery of America" by Christopher Columbus. It would be called the Columbian Exposition and it would open in 1893.

Numerous American cities vied for the right to host the event, and the candidates at last boiled down to New York and Chicago. New Yorkers and much of the East Coast establishment scoffed at the notion that Chicago was anywhere near civilized enough to be the host city. They imagined the midwest metropolis as a frontier town prowled by hicks and louts, and comprised mostly of docks, factories, and stockyards. As those critics should have anticipated, their attitude only caused a diversity of Chicago civic and business forces to coalesce into a formidable entity that was determined to bring the fair to their city.

Architect Daniel Burnham was chosen to lead the project and carry the battle to New York City. In presenting Chicago's proposal to the U.S. House of Representatives, Burnham made the point that Chicagoans were more about substance than style, but that given the chance, Chicago could do style like no other city. On the eighth ballot, Chicago won the opportunity to show what it could do.

The Exposition was to be a celebration of art and industry, and of the products of soil, sea, and mine. Chicago mayor Hempstead Washburne appointed Charles Deere to be one of two fair commissioners to represent the state of Illinois. On hand as secretary of the commission was Benjamin Butterworth, the father of William

Butterworth. William, who married Charles Deere's daughter Katherine in 1892, would one day succeed Charles Deere as president of Deere & Company.

Architect Daniel Burnham entertained many proposals for a "trade-mark" feature for the Exposition, something to rival the Eiffel Tower. Burnham had his own ideas and ignored a proposal for a vertical merry-go-round from a young designer named Ferris. When it seemed that Ferris's "wheel" was the only big-ticket option that could be completed in time, Burnham relented. Ferris constructed a 250-foot-diameter wheel in time for the opening and operated it successfully throughout the fair. People loved it. The Eiffel Tower had been eclipsed.

Charles Deere served as one of two Illinois commissioners to the 1893 World's Columbian Exposition, which commemorated the 400th anniversary of Columbus's landing in America. The sprawling fair of commerce, science, and culture occupied more than 600 acres in Chicago. During the fair's six-month run, exhibits from 46 nations drew more than 25 million visitors. Deere & Company displayed its wide variety of plows at a large exhibit in the huge agricultural building and annex, which covered 15 acres. Among the other attractions in the building: an eleven-ton block of Canadian cheese and a 50-ton French-chocolate statue of Columbus.

COMPETITION, TRUSTS, AND WAR

Until the invention of the gasoline engine, no mechanical device had a greater impact on agriculture than the Appleby Automatic Knotter of 1883. This amazing device turned the grain reaper into a binder, saving thousands of man-hours per harvest. In just a few years, the resultant release of manpower caused a major shift in America's rural-urban population, as fewer workers were needed in the nation's agricultural areas. At the time, binders (or harvesters, as they came to be called) were sold by quite a few companies. Roving sales agents who scrapped over every sale personified an intense competition. Prices fell and sales costs rose. The farmers benefited, but the manufacturers soon tired of battling for very little profit.

Economic recession in the 1890s prompted whole industries to form combines, called trusts, to eliminate competition and control the public's buying. The big players in the agricultural-implement business were squeezed into following suit. In 1891, Canadians Massey and Harris became partners. C. H. McCormick, William Deering, and five smaller outfits came together in 1902 to form International Harvester. Case and John Deere reorganized themselves and acquired smaller companies to remain competitive and avoid unwanted takeovers. Railroads and banking houses also covered themselves with protective trusts.

Following the 1901 assassination of William McKinley, Theodore Roosevelt became president of the United States. Roosevelt was a man of righteous zeal as well as great energy. In his eyes, the "trust problem" had to be resolved. The young president invoked the Sherman Anti-Trust Act, which had been on the books since 1890. International Harvester became a primary target, since five of the companies from which

I-H had been formed had manufactured binders, mowers, and rakes. This, the government maintained, amounted to restraint of trade. Harvester was ordered to dissolve.

I-H appealed to the U.S. Supreme Court. Legal battling dragged on until the United States became involved in the Great War (World War I) in 1917. The government realized that if it won the case against Harvester, the seven other cases pending would automatically follow suit. The result could be the total disruption of war production. Government lawyers were granted an indefinite postpone-

International Harvester introduced the Titan Model D in 1910; this is a 1911 example. The behemoth was named after the figure in Greek mythology of "gigantic size and enormous power." All early I-H models fell into this category but would later decrease in size as the market demanded smaller machines that were easier to operate.

ment, but Harvester, not wanting this threat hanging over it, sought a settlement. In a 1918 consent decree, Harvester agreed to divest itself of the Osborn, Champion, and Milwaukee lines of harvesting machines, and eliminate dual McCormick and Deering dealerships.

STEAM POWER

In 1910, just two years before Deere managers began seriously considering the tractor business, this North Dakota farmer used a massive Minneapolis steamer engine to pull his 14-bottom John Deere plow. By focusing on the quality and features of its plows and other implements, Deere had left the self-propelled field entirely to others. But in the face of growing competitive threats from International Harvester, Deere soon turned its eyes and ambition to tractors.

While agricultural manufacturers adjusted to the changing realities of the marketplace with improvements to their harvesters, harrowers, and other implements, simultaneous work went on in the area of motive power. The first breakthrough, the farm steam engine, resulted directly from progress made in railroad engines. By the mid-19th century, railroads were a boom industry swept along by rapidly developing technology. The Westinghouse air brake and the Janney safety car coupling were enormous improvements to rolling stock, and encouraged further growth. By 1848 the United States had 6,000 miles of track, with another 2,000 miles added each year.

The steam engine was king of the rails, and soon became a presence on America's farms, as well. With the development of the threshing machine, the horse was fast becoming insufficient as a power source. Hence, it was mostly thresher manufacturers that began building portable steam engines for the farm. Such engines were introduced by Archambault in 1849, by Hoard and Bradford in 1850, by Case and Gaar Scott in 1869, and by Rumely in 1872.

In the West, the Daniel Best Agricultural Works built combines (traveling combined harvesters) designed to be pulled by as many as 40 horses. Ground wheels drove the threshing and separating mechanisms. The machines were enormous and potentially dangerous, too, because almost anything could spook such a collection of horses into an uncontrollable runaway. Besides the obvious danger to farmers and anybody else who might happen to be nearby, the machines themselves would be ruined when the ground wheel drove the mechanisms to destruction.

Daniel Best sold his first steam tractor in 1889. A steam-powered combine followed (literally) that same year: Steam from the tractor was piped back through a high-pressure hose to power the combine mechanisms. Because this was a dramatically efficient way to cultivate land, horse-powered treadmills and rotary sweeps began to disappear from the scene. Still, for farmers to make the shift from horses to steam engines involved considerable expense.

Farms then were much smaller, on the average, than those of today, especially east of the Great Plains. In the days before the turn of the 20th century, farming was mostly a self-sufficiency occupation: Farmers grew for their own needs. Besides staples like wheat, corn, potatoes, and hay, small farms raised sheep for meat and wool, cows for milk products and meat, and chickens for eggs and meat. Fruit was grown, and the typical farm usually had a large vegetable garden. In the North, maples were tapped for syrup and sugar. Hunting, fishing, and trapping provided food and skins. The women canned, spun, sewed, baked, churned, and made candles and soap. Excess crops were cashed-out to get the few necessities of life not produced on the farm. It's only natural that one would not want to risk thousands of dollars on a machine that could be used only at threshing time. And if it failed, where would the horsepower come from?

Have Thresher, Will Travel

For most farmers who had questions about thresher technology, the custom thresherman was the answer. Often, the thresherman was the son of a farmer with too many sons to divide the farm between. The most mechanically inclined of the boys would buy a threshing outfit and take to the roads at harvest time. He'd often end the season hundreds of miles away, adequately paid and ready to load his rig on a flatcar for a ride home.

During their heyday, as many as 75,000 custom thresher engineers covered America's farm regions. They followed the harvest from south to north, sacrificing a normal home life and often sleeping beneath their machines. They came back year after year, lured not just by financial reward but by their love of machinery, the satisfaction of a job well done, the excellent food served by the farm women, and by the high esteem of the farm lads.

Steam power had an almost mystical charm for those associated with it (and still does today). Anybody thus infected couldn't ignore a train whistling for a distant grade crossing, or the chuckling sound of a steam engine operating a thresher or a sawmill. Many years later, retired oldsters tried to explain their fascination to the younger generation, but explanation was unnecessary or ineffective. You either had the steam bug, or you were immune.

As the 19th century shaded into the 20th, the success or failure of a small American farm could depend upon the availability of a strong tractor with a belt drive.

Even before Deere & Company entered the tractor business, its presence in the lives of Moline's citizens was enormous. The John Deere School (*above*) is seen as it appeared in 1902. A notation that accompanies the photo says that the school was located "on South side of 17th Ave. at about 10th Street." Today, with a modified name, the John Deere Middle School is on 11th Street, south of 19th Avenue. The image below shows Deere School students on campus in 1943; the school (with some cosmetic changes) is in the left background.

HERE COMES INTERNAL COMBUSTION

At the turn of the 20th century, developed nations enjoyed an explosion of technological advances with an impact similar to the one that characterized the turn of the 21st century. Many Americans of the 1890s had fought in the Civil War. They subsequently witnessed the replacement of frontier log cabins and stumpy fields with prosperous farms dominated by well-groomed fields, frame houses, and huge barns. Barely discernible wagon ruts across the prairies were replaced by railroads. In the lifetimes of Civil War veterans, the telegraph, telephone, electric light, phonograph, and safety bicycle were invented. And that was just the beginning. As the internal combustion engine was developed, the automobile, the tractor, and the airplane followed.

Development of the gasoline tractor progressed in stages. First, stationary engines were mounted to skids to make them portable.

The benefits of farm steam engines were sometimes tempered by the danger of fires or explosions. But to thresherman and inventor John Froelich, those problems presented a great opportunity. In the summer of 1892 he combined a Van Duzen gasoline engine with a Robinson chassis to create the Froelich tractor. During a subsequent seven-week period of traveling "demonstration" threshing, Froelich left little doubt that he was showing the future of tractor technology. His design ultimately led to the development of the famed Waterloo Boy tractor.

Wheels were added, then a drive mechanism, and then a means of steering. Finally, with the addition of a drawbar that allowed the hookup of implements, the concept was complete. The first such tractor was built by the Charter Gas Engine Company of Sterling, Illinois. John Charter's patent, issued in 1889, was based on the use of liquid fuel (gasoline). Prior to that, engines used everything from natural gas to coal dust for fuel. But the need for lubricants in the machine age encouraged growth of the petroleum industry. When petroleum was refined to make oil and grease, a highly volatile byproduct was distilled. It was called gasoline. Because of its flammability it became a glut on the market, but with the acceptance of gasoline as an engine fuel, both the petroleum and engine businesses took off. When Charter mounted his gasoline engine on the running gear of a Rumely steam engine, he was able to sell six of these machines,

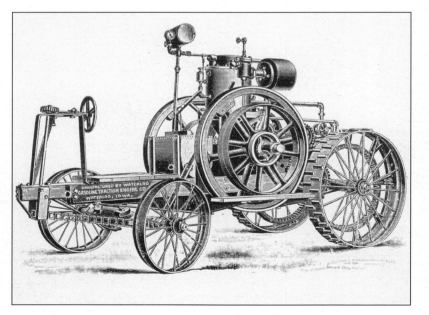

even though they were unable to back up.

The first tractor able to propel itself both backward and forward was the 1892 Froelich. John Froelich, of Froelich, Iowa, mounted a Van Duzen gasoline engine on a Robinson steam engine frame. He devised his own drive and steering systems. Froelich took his machine and his Case thresher on a custom tour across South Dakota. As he traveled, he pulled the thresher from place to place with the tractor. Once set up, he powered the thresher by means of a flat belt connected to the engine's flywheel. In the 52-day run, Froelich threshed some 72,000 bushels of small grain. Later, Froelich joined with venture capitalists to form the Waterloo Gasoline Traction Engine Company of Waterloo, Iowa. This company later built the Waterloo Boy tractor.

Charles Hart and Charles Parr founded the Hart-Parr Gasoline Engine Company in 1897 while they both were engineering students at the University of Wisconsin at Madison. When they had trouble raising capital for expansion, they moved their operations to Charles City, Iowa, Hart's hometown. Since his family was well known there, the local banks provided the cash necessary for the pair of youngsters to venture into the tractor business. Hart-Parr No. 1 was completed in 1901 and sold in 1902. An improved version, Hart-Parr No. 2, was completed late in 1902. Fifteen examples of the Model No. 2 were delivered in 1903. By 1905, Hart and Parr had established the only business in America devoted exclusively to tractor manufacturing, and had in fact, in their advertising, coined the word "tractor." Previously, these devices were called "traction engines." By 1907, one-third of all tractors in the world (about 600) were Hart-Parr machines.

PRODUCTS OF DEERE & COMPANY

Following the Civil War, the Deere product line consisted of plows, planters, harrows, cultivators, buggies, and wagons. This line was maintained until the turn of the century, with many varieties of each main type added as the 19th century wound down. The only all-new Deere line during this period was a bicycle, inspired by the cycling craze that swept the country in 1896 and 1897. A pair of executives at Deere's Minneapolis branch, C. C. Webber and Charles Velie, aggressively brought Deere into the bicycle business.

Deere & Company issued periodic catalogs and price lists. Individual items grew from a total of 31 in 1866 to 287 separate and distinct models of single-blade walking plows in the year 1900, plus a similar variety of the other products. The biggest seller, however, was the sulky, or riding plow. It came in single-, two-, four-, and even six-bottom versions. The heavy six-bottom plow, with no seat for the operator, was obviously designed for steam-traction engine power. Deere had earlier experimented with the Fawkes steam engine without much success, but now was actively pursuing the steam-plow business.

Everywhere you looked in the 1890s somebody new was riding— and making—bicycles. The Deere Model "A" bicycle sold for a hefty $50 in the late 1890s—equivalent to at least $500 today. Deere bought the majority of the bike's components from other manufacturers and bolted them to a proprietary 24-inch frame (22- and 26-inchers could be had, as well), with nickeled, drop-forged forks. The standard enamel was "Brewster green" but maroon and black also were available. The appearance of the automobile in 1903 caused a huge reduction in bicycle sales, and even the handsome, high-quality Deere two-wheeler proved to be short-lived.

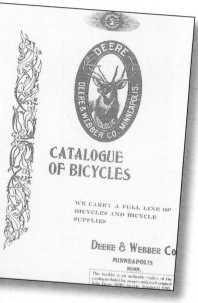

In addition to supporting parades (*right*) and other Moline, Illinois, events, John Deere cofounded one of the city's first banks, served as mayor of the city, and donated land to the city for a market square. The family's central involvement in the history of Moline is evident today in the form of the massive John Deere Pavilion, as well as historic sites, including several Deere family homes.

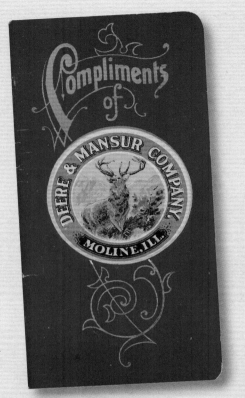

Marketing that Mattered

Effective business promotion helps move product. If the promotion has some practical value, it breeds goodwill, too. For many years, Deere has offered complimentary logbooks to farmers, who use the pocket-size publications to keep track of finances, crop and livestock productivity, land rotation, labor hours, and maintenance schedules for tractors and other agricultural machinery. A typical log also includes useful charts of weights and measures. The one seen here, with a Deere & Mansur logo, dates from about 1890. Deere & Mansur was consolidated into the larger Deere & Company by 1911.

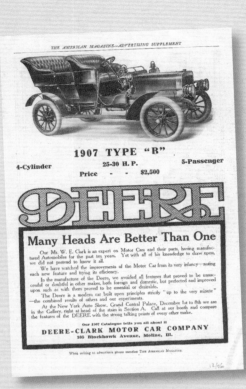

1907 TYPE "B"

4-Cylinder   25-30 H. P.   5-Passenger
Price - - $2,500

DEERE

Many Heads Are Better Than One

Our Mr. W. E. Clark is an expert on Motor Cars and their parts, having manufactured Automobiles for the past ten years. Yet with all of his knowledge to draw upon, we did not pretend to know it all.

We have watched the improvements of the Motor Car from its very infancy—noting each new feature and trying its efficiency.

In the manufacture of the Deere, we avoided all features that proved to be unsuccessful or doubtful in other makes, both foreign and domestic, but perfected and improved upon such as with them proved to be essential or desirable.

The Deere is a modern car built upon principles strictly "up to the very minute"—the combined results of others and our experiments.

At the New York Auto Show, Grand Central Palace, December 1st to 8th we are in the Gallery, right at head of the stairs in Section A. Call at our booth and compare the features of the DEERE with the strong talking points of every other make.

Our 1907 Catalogue tells you all about it

DEERE-CLARK MOTOR CAR COMPANY
105 Blackhawk Avenue, Moline, Ill.

When writing to advertisers please mention THE AMERICAN MAGAZINE.

The five-passenger, 1907 Deere Type B touring car was one of only about 100 automobiles produced and sold by the Deere-Clark Motor Company. The firm operated from 1905 to 1907, falling victim to problems with labor strikes, cash flow, and production schedules. Finally, bankruptcy was unavoidable. The Type B had a water-cooled, four-cylinder engine, steel frame, and a wooden body. Standard equipment included battery ignition, three oil lamps, two gas lamps, a clock, a horn, and other luxuries. The car sold for $2,500 (plus $125 for an optional top) and was well received for its mechanical attributes. Deere-Clark's collapse aside, consumer interest in road travel made it clear that motorcars were the next big thing. Indeed, in 1908 Henry Ford made history with his brand new Model T.

The Butterworth Era

Seventy-year-old Charles Deere died on October 29, 1907. He had been with the company for 54 years. The board created a statement conveying their affection and admiration for him. It read in part: "We, who were his associates for many years thus record our regard and testify to his simple, strong and manly character, and to his sterling worth."

Charles Deere had brought his son-in-law, a lawyer named William Butterworth, into the company after Butterworth married Katherine Deere in 1892. As Charles's health began to fade, Butterworth was given more and more daily control. Although Charles Deere was chief executive officer, Butterworth, who was company treasurer, expanded his purview to more than just the financial realm, and was, in fact, general manager by the time Charles died. Shortly after, he was named CEO.

In 1892, six years after John Deere's death, William Butterworth of Ohio married Deere's daughter Katherine. That same year, he joined Deere & Company as an assistant buyer. He rose to the office of company treasurer by 1897 and became company president when Charles Deere died in 1907. Over the 29 years that followed, Butterworth guided Deere & Company's transformation from plow and wagon maker into tractor-business giant.

JOHN DEERE PLOW CO.

CONDITIONS OF SALE.—PAYMENT TO BE MADE BY DRAFT ON DALLAS OR NEW YORK, MONEY ORDER OR CURRENCY PREPAID. CHECKS ON OTHER POINTS WILL BE CREDITED LESS COST OF COLLECTION. SHORTAGE CLAIMS MUST BE MADE WITHIN FIVE DAYS AFTER RECEIPT OF GOODS. GOODS IN TRANSIT AT CONSIGNEE'S RISK.

No. 2633
ORDERED BY   L   DATE ORDER   11/13   DALLAS, TEX..   11/14/08
SHIPPED BY   Katy   SOLD TO M   Belford Lbr. Co.
BACK CHARGES   2.50   Georgetown, Texas.
SHIPPING TICKET   G 17917   PAYABLE NET   60 Days   (OR LESS   % IF PAID WITHIN TEN DAYS FROM DATE OF INVOICE.)

1 # 104 Cart Red   24.00

To You, Weir, Texas.

Carts and wagons were key parts of the Deere business between the 1880s and the advent of the company's tractor business. This November 14, 1908, bill of sale records the purchase of a red utility cart by a Georgetown, Texas, lumber company. After selling other companies' wagons for a generation, from 1907 to 1911 Deere bought three wagon manufacturers: Fort Smith, Moline, and Davenport.

Tentative Tractor Steps

In the 1880s, the Otto-cycle (4-cycle) engine began to supplant the steam engine as a source of non-animal farm power. The last decade of the 19th century saw a large crop of stationary one-cylinder "gas engines." In 1902, Hart and Parr sold the first regular-production gas tractor and became the fathers of a great industry. The upstart Henry Ford joined automobile pioneers, Duryea, Benz, Olds, and Daimler. By 1900, 8,000 auto buggies toodled on America's streets and roads. But eight million horses still labored across the country.

Hart-Parr was one of 300 U.S. companies making tractors in the early days of mechanized farming. Established in 1907, Hart-Parr devoted itself to tractors only (like this 60-horse Hart-Parr 60 from about 1910), and didn't get led astray by the myriad implements and accessories offered at the time by other manufacturers.

By 1912, when Deere reorganized itself into an integrated manufacturing and sales company, it was already fighting a pitched battle with International Harvester for dominance in the farm equipment industry. While Deere managers still were not fully committed to the idea of building Deere-brand tractors, they had no such qualms about harvesters. Here, five Deere harvesters are pulled by a 10-ton Big Four tractor, which Deere also sold through its catalog at the time.

TRIAL AND ERROR

From approximately 1850 to 1950, the farm machinery business saw almost constant, dramatic change. Each new invention spawned another. Thus, reapers begat threshing machines, and when threshing machines required more power than could be conveniently supplied by horses, farm steam engines were developed. The portable steam engine led to the traction engine, which was followed by the internal combustion traction engine.

Agricultural historian R. B. Gray credits Obed Hussey with the invention of the steam engine plow in 1855. Hussey, from Baltimore, is remembered as the second inventor to patent a reaper, much to the chagrin of Cyrus Hall McCormick, who was third. (The first reaper patent was awarded to Briton Joseph Boyce in 1799.) Hussey's reaper was, however, more successful than his steam plow.

Three years later, a Mr. J. W. Fawkes, of Christiana, Pennsylvania, introduced a more successful steam-plowing outfit at the Illinois State Fair. Based on a vertical boiler steam engine of 30 horsepower, it pulled a mounted

Although built by the Gas Traction Company of Minneapolis, the 19,000-pound Big Four showed up in two Deere-products catalogs in 1912. Because it was common at the time for dealers to sell products produced by other makers, this Big Four tractor, sporting coloration that hinted it was a Deere product, wasn't as unlikely a catalog item as you might think.

The 12–24 tractor, a 1914 product of the Velie Motors Corporation, was used by Deere branch houses of the time to pull Deere plows during demonstrations for customers. Although the impact of the Velie 12–24 itself was limited, a flesh-and-blood Velie played a key role in prompting Deere to build its own tractor before the market was dominated by competitors. Willard Lamb Velie, a son of John Deere's son-in-law Stephen H. Velie, and supplier of the 12–24 tractors Deere was using, openly challenged the Deere board to either pursue tractors "whole-heartedly, or dismiss the tractor matter as inconsequential and immaterial."

six-bottom makeshift plow with a "power-lift." Observing that the plow was the rig's weak point, John Deere contacted Fawkes, and over the summer of 1859 developed an eight-bottom steel plow compatible with Fawkes's invention. In the fall of 1859, the rig was entered in the U.S. Agricultural Contest in Chicago, where it won the Gold Medal. Despite this victory, the great, ungainly Fawkes engine did not prove serviceable, and none were sold to the public. Nevertheless, Deere & Company had its first taste of the tractor business.

In the early 1900s the giants of the tractor and implement industry were International Harvester, J. I. Case, and Massey-Harris; together, they were known as the "Long-line" companies. Deere, with its five lines, was looking forward, but was not yet in their league. By 1906, International Harvester was making a gasoline traction engine. Case had been a leader in the steam engine business since 1892.

Massey-Harris of Canada delayed entry into the tractor business until 1918.

Because Deere provided plows for tractor companies, William Butterworth was reluctant to enter the tractor business for fear of alienating those companies, who might turn their backs on Deere implements. Some customers assumed that Deere was in the tractor business long before it really was. The Gas Traction Company of Minneapolis manufactured one of the best tractors in the country—the 19,000-pound monster known as the Big Four. In 1910, the Big Four appeared in Deere & Company catalogs, with the implication that it was a Deere product. At the very least, the Big Four pointed to Deere's future: In one of the catalog's color spreads, the tractor's predominant color is a green very close to that adopted later for genuine John Deere tractors.

The arrangement with the Gas Traction Company did not last long, but did give Deere a taste of what having its own tractor would be

like. In 1912, the Deere board assigned staff engineer C. H. Melvin to build a tractor. This came to be known as the "Melvin" tractor, a three-wheeled affair with the unpowered single wheel reserved for steering. The machine had two operator seats facing in opposite directions with the steering wheel on a vertical pedestal between. For plowing, the two drive wheels were forward, with three plows mounted underneath. For pulling attached loads, the operator faced the other direction with the unpowered (steering) wheel going first. The Melvin was a gallant effort, but unreliability and lack of traction caused the board to abandon the project.

A more credible attempt followed: the Velie 12–24 of 1916, which was developed by engineer Stephen Velie—a Deere board member and John Deere's son-in-law. Velie's 12–24, which he called the "Biltwell," was created with little help from the parent company, and was used primarily by the branch houses to demonstrate Deere plows.

A BOY FROM WATERLOO

Until 1914, outsized tractors that weighed as much as 15 tons were the industry norm. Many farmers simply had no way to use such monsters. Deere directors commissioned yet another board member, Joseph Dain, to develop a small tractor that would sell for around $700. By early 1915, Dain had a prototype ready for the board to see. It was a three-wheel, all-wheel-drive machine with a four-cylinder Waukesha engine.

By the end of the year, six to ten prototype tractors were proving their practicality in the field. After a year of testing, and with a more powerful McVicker engine, 100 were built for sale.

Unfortunately, Dain died in 1917, and much of the push for the tractor dwindled. While the so-called Deere "Dain" was ahead of its time and an engineering success, the target price of $700 had ballooned to $1,700. In the meantime,

Frank Silloway, Deere head of sales, learned that the Waterloo Gasoline Engine Company was for sale. Their 25-horsepower tractor, the two-cylinder Waterloo Boy, was selling well at just $850. Even better from Silloway's point of view was that Waterloo was available for $2,350,000, with a factory already turning out about a hundred tractors a week.

The Waterloo Boy was a direct descendant of the world's first truly

The first Deere tractor to go into production was developed by board member and engineer Joseph Dain, Sr. The three-wheeled Dain machine improved on the short-lived Melvin design—for one thing, power went to all three of the Dain's wheels, instead of to only one. By the time the first 50 tractors were sold, Dain had died, and Deere managers had their eyes on a different prize altogether: the Waterloo Traction Engine Company, which produced Waterloo Boy tractors, and sold them by the thousands.

successful internal combustion traction engine developed by John Froelich in 1892. Later that year, Froelich joined with others to form the Waterloo Gasoline Traction Engine Company. Four tractors of the Froelich design were built and two were sold to customers. Neither proved satisfactory and both were returned to the company. To

Motivated in large part by the commercial success of the Waterloo Boy Model N tractor, Deere bought the Waterloo Gasoline Engine Company in 1918. The N ran with a two-cylinder engine and could be steered like an automobile. The tractor burned kerosene but, for cold starts, had small tanks of gasoline along the fenders. Deere continued to improve the tractor's design for several years before introducing the Deere Model D in the mid-1920s.

generate cash flow, the company developed stationary engines, which they sold successfully. When the company reorganized in 1895, the word "Traction" was dropped from the corporate name. With that, Mr. Froelich left the company. By 1906, six stationary-engine models were in production, with the trade name "Waterloo Boy." In 1911, Moline resident A. B. Parkhurst joined the Waterloo Boy Gasoline Engine Company, bringing with him three tractors of his own design, each with a two-cylinder engine.

From 1911 to 1914, Waterloo Boy tried many variations on the two-cylinder theme. Model designations and serial numbers from this period are confusing because test vehicles were often rebuilt and redesignated. Finally, in early 1914, the design of the Waterloo Boy Model R tractor was set: It was to be a four-wheel, rear-wheel-drive machine,

with one forward speed and one reverse. The engine was a two-cylinder, four-cycle overhead valve type, with a bore and stroke of 6.6"×7" giving a displacement of 333 cubic inches. Operating speed was 750 rpm, which produced 12 to 24 horsepower at the belt. The drawbar rating was 12.

The Model R was available in 13 iterations, A through M, until 1918, when Waterloo was purchased by Deere & Company. Style M, which was renamed Model N, was introduced in 1917 and remained in production until 1924, when it was replaced by the first mass-produced two-cylinder tractor bearing the John Deere name: the venerable Model D. (The four-cylinder Dain also bore the John Deere name, but most were bought back and destroyed by Deere after the purchase of Waterloo Boy.)

The success of the Waterloo Boy Model R tractor—the popular companion product to Waterloo's Model N—caught the attention of Deere directors who were working to develop their own tractor. Both models had two-cylinder, kerosene-burning engines. The R, introduced in 1914, sold for $985 when Deere bought the Waterloo line in 1918. Following its familiar pattern of acquisition and improvement, Deere incorporated the best qualities of the Waterloo models into the tractor design that became the famed Deere Model D six years later.

DEERE ANSWERS THE COMPETITION

In 1900, Deere's chief competitor was U.S.–based Oliver Chilled Plow Works and Canada's Cockshutt. After the 1902 formation of International Harvester, Deere management became concerned not simply about product competition but about I-H winning the hearts and minds of the independent dealers. If this happened Deere would be forced out of business or end up bent under Harvester's control. William Butterworth was probably most concerned by the latter prospect.

Butterworth led the Deere board to a policy of aggressive defense. To prevent being swallowed up by I-H, they would have to become too large a bite. Yet the tractor part of the business still presented a quandary. Deere management struggled with the question of whether to enter the tractor market or to continue to tailor their implements to the tractors of others. To enter would alienate some of their best customers; not to enter would leave Harvester with a big advantage.

At first, Deere & Company took a hands-off, self-effacing stance, emphasizing to dealers and farmers that the two companies' product lines didn't compete. International Harvester made harvesters and mowers and the like; Deere made plows and cultivators. Many independent dealers sold both lines of implements. In 1906, however, I-H took a more hostile competitive posture by encouraging exclusivity of its product lines at the dealer level. Harvester also began to market farm wagons, gas engines, and manure spreaders—sometimes on the

An August 1916 tractor show in Fremont, Nebraska, drew crowds from far and wide, including a vacationing Henry Ford and an entourage of two dozen of his employees. During the event, 50 companies displayed more than 250 tractors, including a novel Deere prototype and several International Harvester Moguls. Deere had not yet committed to full production of tractors, but the success of Fremont exhibitors, who generated sales valued at $1.3 million, was a sign that Deere needed to move quickly or risk being left behind in the tractor business.

Just prior to World War I, after John Deere signaled it would expand its product line by building a harvester factory in East Moline, Illinois, harvester icon International Harvester responded by acquiring two plow manufacturers. By 1915, the battle between the two companies was fierce, as shown by this John Deere grain binder displayed outside an I-H factory, festooned with handwritten signs detailing alleged "deficiencies" in the Deere design.

From 1915 to 1922, International Harvester built nearly 90,000 Titan tractors, including this 1917 10–20 model. The very powerful, 5,700-pound Titan burned kerosene and sold for $700, making it a formidable competitor for Deere's early tractor lines.

sly—after purchasing the companies that made them.

To defend itself from further monopolization of the marketplace by Harvester, Deere acquired the Fort Smith Wagon Company in 1907, just before the death of Charles Deere. Later that year, the John Deere Plow Co., Ltd., was formed in Canada. Finally, on January 6, 1910, Deere directors issued a directive reorganizing the company into a more consolidated entity. The same written decision included the statement that Deere & Company would enter the harvester business.

In 1911, Deere undertook an aggressive policy of acquisition. Companies were brought beneath the corporate umbrella, and Deere was shortly making shellers, elevators, spreaders, and hay-making equipment. Some of these companies were moved to the Moline area. Finally, in 1912, ground was broken in East Moline for a new "harvester" plant. Now the cat was out of the bag. Deere was challenging International Harvester head-on. It was almost a foregone conclusion that tractors would be next. The 1918 purchase of Waterloo Boy, and the commercial success of those early Deere tractors, began the metamorphosis by which Deere became the tractor-making giant we know today.

JOHN DEERE MEMORABILIA

Deere's "Oblique Selection" seed-planting system promised corn growers thorough, evenly spaced coverage. This ad is from about 1914.

Deere branch houses commissioned product from the Moline Wagon Company, an organization that merged with Deere in 1910 and became John Deere Wagon Works. In this image from November 21, 1908, farmer J. H. Mathis takes a breather while perched on his Deere wagon.

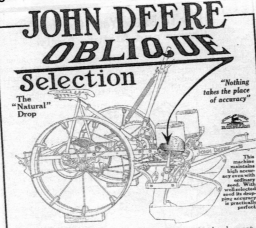

JOHN DEERE OBLIQUE Selection

The "Natural" Drop

"Nothing takes the place of accuracy"

This machine maintains high accuracy even with ordinary seed. With well-selected seed its dropping accuracy is practically perfect.

MAKE your corn ground pay you with the largest possible yield, by making sure that the required number of kernels is in each hill.

The seed must be there. Every "miss" means just that much thinner stand. No amount of cultivation can make up for inaccurate planting. The loss of one ear from every hundred hills costs you the price of one bushel per acre.

For years the John Deere Company has concentrated on accuracy in planting devices. The "Oblique Selection" solves the problem.

The results of its use are so profitable that many corn growers have discarded the best of previous machines. It is as far ahead of the old Edge Drop as it was ahead of the round hole plate.

Make your spring planting the start of your biggest corn crop by accurate dropping. The John Deere "Oblique Selection" will do it. The machine is a splendid investment.

Free Book Gives Valuable Corn Facts

Write us today for free booklet "More and Better Corn." It tells you why the average yield for the United States is only 25 bushels an acre, whereas better methods have produced 125, 175 and even 255 bushels per acre in places. It also describes and illustrates the John Deere "Oblique Selection" Corn Planter. The book to ask for is No. D-110.

JOHN DEERE, MOLINE, ILLINOIS

A Deere & Webber bicycle promotional piece.

A finely preserved 1896 Deere & Webber bicycle.

Manufacturer nameplates from the 1896 Deere & Webber bicycle line.

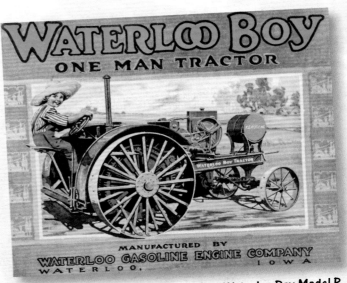

This is a 1916 dealership poster for the Waterloo Boy Model R. The happy boy, whose presence suggested that the tractor was easy to operate (a stretch, perhaps), appeared in many Waterloo Boy posters and ads. The Model R was discontinued late in 1917, and was replaced by the Waterloo N.

In 1912, a representative of Iowa agricultural-equipment dealer W. W. Hubbert Co. wrote to Deere's Omaha branch house to describe a leak in the Simplex bowl component of a customer's grain mill. The problem was almost certainly addressed to everybody's satisfaction.

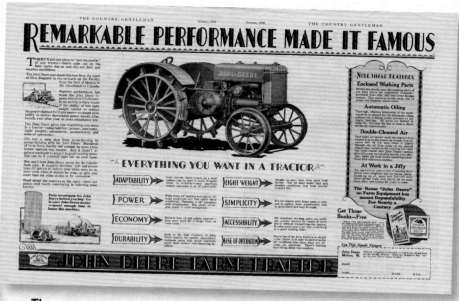

The numerous advantages of Deere's Model D Two-Speed are enumerated in this 1929 ad from *The Country Gentleman*. This model was an improvement over the D of 1923–25, and would get even better with changes adopted in 1931.

Deere's Cornet Band was organized in 1884. The employee musicians played at fairs and other Moline events.

"See us about any implements you need." That was the invitation offered to recipients of this sample issue of the John Deere magazine, *The Furrow*, in 1912.

BRITAIN, FORD, AND WORLD WAR I

On June 28, 1914, in the city of Sarajevo, a Serbian nationalist named Gavrilo Princip assassinated the Crown Prince of Austria, Archduke Franz Ferdinand. Austria-Hungary presumed that Serbia harbored a terrorist organization, and it declared war. Because of an absurd network of mutual defense treaties, within about six weeks Germany, Russia, France, Turkey, Bulgaria, Belgium, and Great Britain were dragged into the conflict, which came to be known as The Great War, later dubbed World War I. The United States felt compelled to enter the fray on April 6, 1917, some three years after hostilities began.

Britain and much of Continental Europe relied on grain imported from America and Russia. British farmers had switched from grain to animal farming when grain prices were low. When the Turkish Navy blockaded the Dardanelles in 1914, the flow of wheat from Russia was effectively shut off. Later, the German U-boat threat curtailed shipments from other sources. The British government reacted by establishing tillage goals for arable acreage. Records indicate that there were just 500 tractors in Britain at that time. The British Board of Agriculture ordered as many tractors as they could get from domestic and U.S. manufacturers. By 1918, about 4,000 Waterloo Boys had been exported to Europe. Most of those that went to England were purchased by London's Overtime Tractor Company, which sold them under the Overtime brand name.

Henry Ford, the automobile magnate, was at the time developing a farm tractor that, he hoped, would do for the farmer what his Model T car had done for the motoring public. To avoid problems with stockholders, he formed a separate, wholly owned family company called Henry Ford & Son. Logically enough, the tractor was called the Fordson. The chief of British Ford, Sir Percival Perry, was on Britain's agricultural board, and pressured Henry Ford to produce tractors for the British food effort. On June 28, 1917, Ford announced an order from the British Ministry of Munitions (MOM) for 6,000 Fordson tractors. Ford's massive production capability was brought to bear, and these tractors were produced in the first few months of 1918. Ominously for Deere and others, Henry Ford was now in the tractor business.

The Waterloo Boy was the machine that put John Deere in the tractor business. Deere purchased Waterloo in 1914, but prior to that, Waterloo often shipped its Model R overseas. Most of those that went to England were immediately given a fresh paint job, decals, and a serial number by L. J. Martin, owner of the Overtime Tractor Company of London. Once refinished, the tractors were sold under the "Overtime" name. They ran on paraffin, a UK equivalent of kerosene. After Deere took over the company, the tractors were rebadged to display the Deere factory green. Between 1914 and 1917, about 8,000 copies of the Waterloo Boy Model R were assembled. To suit farmers' needs, the model was offered in 13 different styles. By 1917, the Boy's design was state of the art, and even its modest 24 horsepower was seen as a marvel in those late days of horse-drawn implements.

THE INFLUENCE OF HENRY FORD

Henry Ford was known as a pioneer in many aspects of the manufacturing industry, and some of his innovations were eventually adopted by other companies. After spending time in meat packing plants, the idea of applying assembly line methods to produce automobiles came to Ford. Utilizing these newly learned techniques, in 1914 he was able to decrease the time required to build a chassis from 12.5 hours down to just 93 minutes.

An innovator as well as a hardheaded pragmatist, automaker Henry Ford revolutionized American industry when he perfected the moving assembly line that allowed cars to be built as they progressed from one workstation to the next. He would bring a similar efficiency to his tractor business.

Perhaps seen as even more radical was Ford's decision to pay workers $5 a day for their labors in 1914. Considering the fact that the average wage of a U.S. worker was 22 cents an hour in the same year, Henry's rate seemed extreme. A worker at Deere's Plow Works took home 31 cents per hour and worked an average of 44 hours per week. The Ford rate was double that of Deere's rate and was implemented to keep workers on the line after they had been trained. Henry saw the value of retaining a trained employee versus having him take a better position elsewhere after spending money to teach him a skill.

Records indicate that Deere continued to pay the same wage from 1914 to 1919, likely due to the radical drop off in business: A deep slump in farm equipment sales for the period did nothing to assist Deere workers or their pay scale. Deere's market share dropped from 15.1 percent to 12.2 percent in the years from 1913 to 1918 as World War I took its toll on every facet of civilian manufacturing. Also feeling the pinch of the wartime downturn, International Harvester paid their employees even less than Deere at the same time. Yet Ford's transformative policies would influence the factory workday and pay rate for generations to come.

DEERE TAKES A CUE FROM FORDSON

The automobile and the American road system developed together, each spurring the growth of the other. As more people experienced the freedom and mobility of automobile travel, the mysteries of the internal combustion engine, driving, and auto repair were gradually understood by increasing numbers of average Americans. Not unlike the boom in personal computer use in the late 20th century, the gasoline age came on suddenly, and with it, the farm tractor.

Henry Ford's Model T revolutionized the production process in 1914, when the moving assembly line was introduced. The price of the car was eventually cut to less than one-half of its original retail. Further, Ford doubled the wages of most of his factory workers to $5 per day, reasoning that these employees would then be able to afford the cars they were building. Ford initiated the eight-hour day at the same time, and because of the visibility of the Ford empire, these measures had a ripple effect on other manufacturers, including Deere.

Fordson tractors were sold in America beginning in 1917. Ford's tremendous production capacity and the Fordson's low price made the new machine immensely popular with small farmers. Deere recognized that the Fordson had a considerably more car-like appearance than its own Waterloo Boy. Thus, a redesign was undertaken that resulted in the venerable John Deere Model D.

Although a carmaker, Henry Ford was keenly attuned to agricultural machinery and its ability to transform and improve people's lives—and make a profit for Henry, too. The first Fordson tractor appeared in 1917 but Henry began developmental work in 1907, when he was photographed aboard what he called his experimental "automobile plow."

When Deere & Company purchased the Waterloo Gasoline Engine Company in 1918, it took control of the Waterloo Boy, a kerosene-powered machine with a good reputation. Advertisements touted the "Long Life" of work purchasers could expect. Deere produced the Waterloo Boy as an exercise in R&D, engineering many refinements that showed up later on the Model D.

The Reign of the Two-Cylinder Tractor

"A four-cylinder engine—not on your life. . . . The John Deere two-cylinder engine has been so outstandingly successful that there is no thought of a change."

—Deere VP and general manager L. A. Rowland in 1937; his words would ring true for another 16 years.

The year 1912 was a watershed for John Deere & Company and the tractor business. Another manufacturer's tractor, the Gas Traction Company's Big Four, appeared in a Deere branch house catalog that year. It was also the year that the John Deere board directed staff engineer C. H. Melvin to construct an experimental tractor. Although some work was done, serious efforts to get Deere into the tractor business weren't made until 1917.

Entry into the tractor business was fraught with difficulties and setbacks all along the way. Yet it was the stress of these times that ultimately strengthened Deere, and positioned it for its eventual domination of the industry.

THE TWO-CYLINDER ENGINE: WHYS AND WHEREFORES

An original characteristic of the John Deere tractor was the odd-sounding two-cylinder engine. It may seem remarkable today, but while Deere's competitors went to "modern" four- and six-cylinder engines, John Deere had bested them all in sales with its two-cylinder "Johnny Popper" engine.

The Waterloo Boy Model N was manufactured from 1917 to 1924. Deere dropped the chain steering for auto steering and replaced bolted frames with riveted units.

THE TWO-CYLINDER ENGINE

From day one, Deere was content building tractors powered by two-cylinder engines. Output and simplicity were adequate and only a handful of buyers complained loudly about wanting more. The success of the company had been based on the twin-cylinder design and few wanted to rock the boat. In a John Deere service bulletin dated February 15, 1937, L. A. Rowland stated "A four-cylinder engine—not on your life." He went on to exclaim that "the John Deere two-cylinder engine has been so outstandingly successful that there is no thought of a change." As the VP and general manager at the time, his words on the subject would be law for another 16 years.

Within the tractor community there were others who used three-, four-, and six-cylinder mills under the hoods, even in the earliest days of motorized farming. Outside the world of crop management a variety of new configurations was being tested and installed into civilian machines. A few luxury automotive names were going the distance with 12- and 16-cylinder engines, but only the rich and famous could afford the vehicles that carried them.

It was 1953 when Deere & Company saw the need to move ahead with plans to create tractors fitted with additional cylinders. Seven years would be needed to bring the plan to fruition, but the 1960 New Generation models with four- and six-cylinder powerplants were welcomed with open arms by loyal John Deere buyers.

John Deere's first tractors had two-cylinder engines, and the company stuck with the relatively simple mechanical design until 1960. The two-cylinder tractor was a fuel-sipper, and its lack of engine belts and pulleys made it simple enough for farmers to make some repairs on their own. Although the postwar economic boom created a busy market for Deere models, pressure for more powerful tractor engines came, ironically, from soldiers returning from duty after having used then state-of-the-art vehicles with four- and six-cylinder engines.

So how did Deere settle on the two-cylinder arrangement? Why did the company stick with it so long, and why did Deere finally discontinue it in favor of more conventional engines in 1960?

Deere inherited a side-by-side horizontal, two-cylinder engine when it acquired the Waterloo Boy Company in 1918. The history of the Waterloo Boy extends to the 19th century. Early internal combustion farm tractors were in many respects modeled after steam engines. Most used four-cycle, single-cylinder engines. Even at a mere ten horsepower, it was impossible to completely balance the up-and-down motion of the piston with rotating counterweights. At 20 horsepower, these single-cylinder tractors tended to hop, or lope. Multi-cylinder engines helped to address the balance problem, and one particular engine type, the horizontally opposed, was virtually self-balancing.

With that configuration, however, physical size became a problem, and unless multiple carburetors were used, the fuel-air vapors tended to condense in the long passages. Two-cycle, two-cylinder engines, with the cylinders arrayed side-by-side but with pistons on opposite crank throws, balanced well and gave even firing. But at this early stage of development, the two-cycle principle had insurmountable technical problems and was soon discarded.

Four-cycle versions of the side-by-side two-cylinder engine, such as those used by John Deere, offered reasonable balance but fired unevenly. The four-cycle, two-cylinder side-by-side arrangement, with the pistons moving in opposite directions, meant the firing occurred at 0 degrees and at 180 degrees of crankshaft rotation. The engine then rotated 540 degrees before firing again. This accounts for the unusual *pop-pop*, pause, *pop-pop* sound (like a double-barreled shotgun; you fire each barrel, reload, and fire again). A heavy flywheel smoothed out the firing to an acceptable level. Besides Waterloo Boy, many early tractor companies adopted this approach to engine balance, including Case, Hart-Parr, and Rumely.

In 1918, Deere & Company abandoned the corporate effort to develop a tractor and bought the Waterloo Boy outfit. With that, Deere was in the tractor business with a well-functioning

> "Drop **all** tractor **expenditures.**"
> —William Butterworth, Deere CEO, 1912

An early Deere General Purpose (GP) tractor was given the cutaway treatment in this Deere handbook for farmers published in around 1932. Parts shown in color were oiled automatically. Deere promised that the GP "replaces horses rapidly and safely."

two-cylinder (horizontal side-by-side) engine.

Other tractor companies dropped their two-cylinders in favor of fours and sixes. Deere found advantages in the horizontal twin, such as better vaporization of kerosene fuel (because of short passages and the proximity to exhaust heat); and, via a crankshaft positioned transversely to the tractor's centerline, the belt pulley was integrated with the clutch on the end of the crankshaft. Also, expensive bevel gears were not required in the power train to the wheels. Further advantages touted by Deere were clutch repair that was possible without splitting the tractor, fewer parts to fail or replace, and safer hand-starting by rolling the exposed flywheel, rather than by twisting a crank and risking a broken wrist.

After about 30 years of nothing but two-cylinder tractors offered in many models (some with vertical engines with crankshafts in line with the tractor's centerline), Deere recognized that the

two-cylinder had reached its limits. As early as 1949, the engine had been adapted to the diesel configuration for Deere's new Model R, which ran with 51 horsepower. Above that, volumetric efficiency, size, and weight dictated more and smaller cylinders, and higher revolutions per minute. For example, the largest two-cylinder diesel, used in the Deere Model 830, had an 8-inch stroke; it displaced 471.5 cubic inches and produced 75.6 horsepower at 1125 rpm. The next-generation four-cylinder diesel of 70 horsepower had a mere five-inch stroke, operated at 2500 rpm, and displaced 271 cubic inches. Nevertheless, farmers loved the exhaust sound given off by the two-cylinder engine. Some thought fondly that the uneven firing of the big, slow two-cylinder mimicked the

beat of the human heart. As if to prove the point, the last two-cylinder Deere tractor, the Model 435 of 1959, had an even-firing supercharged two-cycle diesel made by General Motors—and made a sound that attracted almost no one!

The Model D was the first tractor produced by John Deere after the company bought Waterloo Boy in 1918. By 1944, 21 years after its introduction, the D carried a stout radiator valance and a full-length cowl (*shown*). Nearly constant changes made it one of Deere's most successful models, and sales were robust until Deere introduced its diesel-powered Model R in 1949. Still, the D remains prized today by collectors.

8-, 12-, AND 16-CYLINDER TRACTOR ENGINES

Before the 1920s, usage of V-8 engines in tractors was limited. By the 1950s, the Funk Manufacturing Company converted some Ford tractors to Ford industrial V-8s, making them the most powerful wheeled farm tractors of the time. It wasn't until the diesel era that V-8 tractors became common. The John Deere Model 8850 of 1982, for example, used a 955-cubic-inch V-8 diesel that cranked out 304 horsepower. Caterpillar used V-12 engines in its large crawlers, and several prototype articulated tractors appeared in the 1970s with 16 cylinders (one achieved with two V-8 engines).

The Fordson and Deere's Model D

Auto magnate Henry Ford introduced his Fordson tractor in 1917. It was lighter and smaller than the Waterloo Boy (which Deere wouldn't purchase until 1918) and ran with a four-cylinder engine. It looked more like a car than a steamer, and sold for much less than the Waterloo Boy. Soon the Fordson was the best-selling tractor in the world. By 1923, Deere's response to the challenge of the Fordson was ready. It was the company's soon-to-be-famous Model D, powered by a modern version of the Waterloo Boy engine covered by a car-like hood. The D had more power and traction than the Fordson and quickly developed a reputation for reliability. Even at a higher price, farmers loved it, and so did industrial and construction concerns who bought the Model DI (for "Industrial") in significant numbers.

Because some executives at Ford Motor Company didn't like Henry Ford's desire to build tractors, Henry formed a new company named Ford and Sons, which was later shortened to Fordson. The first Fordson, the Model F, was produced in 1917; the one seen here is the 1918 model.

Deere introduced the Model D in 1923. It was created to supplant the Waterloo Boy Model N, and the two models were sold simultaneously for one year. The D was powered by a fuel-sipping two-cylinder motor. The tractor's stout configuration made it a jack-of-all-trades that could be used for a variety of farming chores. The D would go on to become the longest-running Deere of all time, with production of various iterations continuing through 1953.

The Coming of the All-Purpose Tractor

Henry Ford's Fordson knocked International Harvester out of first place in tractor sales. I-H responded with a unique redesign of its tractor concept—the Farmall. The Farmall was designed to do all the work around the farm that a horse could do, as well as that of a conventional tractor. It had a belt pulley as well as a "power-take-off" (PTO) shaft in the rear to power harvesters; it had a mounted cultivator; and was built with enough clearance between the ground and the rear axle to allow corn cultivation well into the growing season. Rather than a conventional front axle, the Farmall's front wheels were close together, to run between rows. The tractor had individual brakes for each rear wheel, which, along with the pivoting front wheel set, allowed the tractor to turn smartly in its own length. The Farmall looked top-heavy, but heavy-weight rear wheels stabilized it and brought increased traction.

The Farmall engine was a four-cylinder that put out some 20 horsepower via a three-speed transmission. Within a few years the Farmall swept I-H back into first place in tractor sales, and by 1928, it had driven the Fordson from the field altogether. (Production of the Fordson resumed in Ireland and England, where it remained a favorite, mostly unchanged, until 1946.)

The Farmall's success encouraged I-H to develop several variations on the theme, both larger and smaller, as well as an improved version of the original. Because the original had no designation other than "Farmall," it has since been referred to as the "Farmall Regular."

International Harvester's Farmall H was easy to start, run, and maintain, making it a natural for farmers with limited time or patience. And as suggested by the 1947 model seen here, the H also was a "styled" machine with good looks. That, coupled with I-H's expansive dealer network (one that was much larger than Deere's), helped make the H America's best-selling farm tractor during 1945–51.

International Harvester's 1924 Farmall tractor took the row-crop market by storm and put pressure on Deere engineers to respond; a decade passed before the Model A and Model B reasserted Deere's tractor leadership.

THE TRACTOR WAR OF 1922

Cyrus McCormick III's 1931 book, *The Century of the Reaper*, recounts one side of a telephone conversation between International Harvester's Springfield, Ohio, office and the Chicago headquarters:

"What? What's that? How much? Two hundred and thirty dollars? Well, I'll be... What'll we do about it? Do? Why damn it all, meet him, of course! We're going to stay in the tractor business. Yes, cut $230. Both models. Yes, both. And, say, listen, make it good! We'll throw in a plow as well."

The words are those of I-H's general manager, Alexander Legge, who was visiting the Springfield works. This telephone conversation came on the eve of the 1922 National Tractor Show in Minneapolis, where Henry Ford had just announced a price cut of $230, bringing the cost of a Fordson to just $395. Most tractor companies spent more than that on just their engines.

Harvester fought back by cutting its tractor prices by $230. I-H machines still cost almost twice as much as the Fordson, but the inclusion of a Parlin Orendorff plow sweetened the deal. The real winner of the "war" was the farmer: The intense competition not only drove prices down, but led to great improvements in interchangeable parts, dealer responsiveness, and tractor performance. Weaker companies were eliminated in this "survival-of-the-fittest" atmosphere. By 1929, only 47 of the original 200 tractor makers were left. One of those survivors, of course, was Deere.

THE **FORDSON** $495
F.O.B. DETROIT
PULLEY AND FENDERS EXTRA

Deere introduced the Model D in 1923, little imagining that the tractor, in unstyled and styled iterations, would enjoy continuous production for 30 years. The spoked flywheel and fabricated front axle identify this 1923 D as one of the first 50 made.

A padded seat that allowed for lower-back support made life with this '48 D considerably easier than with earlier models.

THE MOLINE UNIVERSAL

The first successful, truly "all-purpose" tractor was the 1915 Moline Universal. This unique machine rode with its engine, transmission, and drive wheels at the front, with a sulky-like unit behind, with a seat and steering wheel. The Universal was articulated, with the hinge just aft of the power unit. A variety of implements could be attached either beneath or behind the trailing unit. The tractor was also unique in that it was the first to be equipped with a starter and lights. Unfortunately, it did not survive the Tractor War of 1922, but its features pointed the way for Deere and others.

The 1915 Moline Universal Model C (*above*) offered practicality as well as a novel drivetrain setup that placed the large drive wheels ahead of the operator. Direction was controlled with an extremely long—and none too precise— steering linkage. Advertisements of the day (*right*) cannily emphasized the "one man" aspect of the Universal—a quality that meant a lot in a time when mechanization had yet to come to most American farms, and when hired hands (who had to be fed and often housed) were needed to perform everyday work. This ad, from 1917, reminded farmers that many able-bodied men were about to be lost to farm work because of America's entry into the Great War.

FINE-TUNING THE GP

Because of the success of Deere's Model D and the challenge of the International Harvester Farmall, in 1928 Deere launched the two-cylinder Model GP. The GP actually started life as the Model C in 1926, but the "C" designation was changed after 110 production units were sold. The GP designation, for "General Purpose," was thought to be a more effective way to answer competition from the Farmall. Further, because of the poor quality of telephone communications at the time, "C" sounded too much like "D," and parts orders had been getting mixed up.

Yet the GP wasn't as successful as Deere had hoped. Rather than having a narrow, or tricycle, front like the Farmall, it had a wide front with an arched front axle. On a Farmall, the driver negotiated the narrow front between two rows while the rear wheels straddled the rows. The mounted cultivator did its job on those two rows. Deere's GP, on the other hand, straddled a center row with its arched front axle, while its rear wheels

Despite the success of its Model D, by the mid-1920s Deere had fierce competition for the row-farming market from International Harvester's hit model, the Farmall. The experimental 1927 Deere Model C, designed to pull a multiple-row planter attachment and perform other tasks, was intended to meet the all-purpose needs of small row farmers. Because of technical challenges and disagreements at Deere over small-tractor technology, only a few Model Cs were produced. However, the effort led to the emergence of the more successful GP model, which Deere produced until the mid-1930s.

straddled three rows that were tended to by the GP's mounted cultivator.

Deere's approach was valid, but farmers never seemed to warm up to the three-row concept, probably because it required too broad a field of view. Also, the GP's arched front axle didn't provide enough clearance for tall crops like corn. And then there was the GP's 312-cubic-inch side-valve engine, which produced a game but frankly inadequate 20 horsepower. Continuous engine improvements were required until 1930, when the engine's displacement was increased to 339 cubic inches and power output, at 25 horsepower, became competitive.

The Model C was released in 1927. It was renamed the GP (General Purpose) a year later because Deere thought that "D" and its previous "C" sounded too much alike. The C/GP was Deere's hope of competing against the new Farmall tractor from International Harvester. It, too, was hailed as a "general purpose" model. The Deere GP would be sold until 1935.

In the 1930s, Deere's Model GP, or "General Purpose," tractor added improved horsepower to the versatility of earlier Deere all-purpose models (note the Ford automobiles that flank the tractor outside a John Deere Farm Equipment building). But by this time the devastating effects of the Great Depression were taking a toll on sales of all tractors. To compete, especially in the growing California market, Deere combined its sales effort in the mid-1930s with Caterpillar, then the leader in tracked-tread tractors.

Deere's GP (General Purpose) tractor came on the scene in 1928 as a do-it-all model designed to meet every need of a contemporary farm operation. It was heavier and more sophisticated than the earlier Model D, but needed repairs in the field more frequently than should have been necessary. Between that and the onset of the Great Depression, GP sales never measured up to Deere's projections. The view from the GP's catbird seat (above, right) was utilitarian, and because comfort wasn't yet a priority, the drilled steel saddle was austere. The access panel was clearly marked to show manufacturer and place of assembly. The GP seen to the right is a late-'29 model, as evidenced by the vertical air intake that protrudes from the cowl.

The GP was named for its general-purpose design that aimed to satisfy all types of farming needs. A two-bottom plow could be pulled behind, giving the GP the ability to plant and cultivate three rows of crops in a single pass. This 1930 model displays the large radiator and stout design that were well suited to rugged duty. The 1930 GP was produced in standard and wide-tread variations; all GPs were fitted with three forward gears and one reverse.

JOHN DEERE
MEMORABILIA

How do you ship Deere equipment? Well, with a shipping tag like this one that dates from sometime in the 1920s. The tag tells us that the item left Deere's Moline, Illinois, plant and shipped through Indianapolis on its way to a John Weis, who lived in Mishawaka, Indiana.

Deere & Webber was the Deere branch house in Minneapolis. The branch was partly owned by Stephen Velie, who was married to one of John Deere's daughters, Emma. Although Velie had been instrumental in Deere's early tractor development, by the beginning of 1918 (the date on this piece of letterhead) the company still wasn't selling tractors of its own. Not quite.

Deere touted the time-saving qualities of its circa 1920 line of disc and other sorts of cultivators. A balance lever that was part of the automatic horse lift helped manage the depth of cultivation, and allowed quick switches to shovel or spring-tooth cultivators.

The Furrow was a Deere & Company farm-information magazine that was distributed free to customers by branch houses and dealerships. This 1928 issue was handed out near Plains, Kansas.

It may look like a general store, but this is the small-parts section in a Deere dealership of the 1920s. The coal- or wood-burning stove kept everybody comfortable during the harsh days of winter.

Deere was a late-comer to the tractor business, and Chicago-based United was just one of many competitors. This ad appeared in the October 1929 issue of *The Country Gentleman.*

This 1928 Deere manure spreader was more simply constructed than many competing products, and was less apt to suffer downtime for repairs. Smooth bearings and a low draft meant that the spreader operated well with just two horses, a considerable savings over the three or four animals that most other spreaders required.

On the cover of this 1921 brochure, a Deere tractor (probably a Waterloo Boy Model N) pulls a disc harrow with a yielding lock. The lock was a coupling device located between the harrow's front and rear sections. By yielding on turns and locking on straightaways, it ensured that the rear discs always cut into the ridges left by the front ones.

SPECIALIZED TRACTORS

Besides the clear move by tractor manufacturers toward two distinct types of tractors in the late 1920s, the "General Purpose" tractor and the "Plowing" tractor (also called the "Standard," "Regular," or "Wheatland" tractor), the period also brought other specialized issues.

The first was the "Orchard Tractor." John Deere's original orchard tractor was based on the GP. Fender skirts covered the rear wheels down to below the hubs, and extended over the flywheel and belt pulley. The exhaust was routed under the tractor, and the seat and steering wheel were conveniently lower than on other Deere machines. Its designation was Model GPO. Several of these were purchased by the Lindeman Brothers of Yakima, Washington, and converted to crawlers (with tank-like treads) for operation in the hilly apple orchards of the Pacific Northwest.

As tractors became more specialized, debate arose over the use of rubber tires rather than cleated, earth-chewing steel wheels. The Model T Ford (the "Farmer's Friend") rolled on rubber tires, of course, prompting the thought, *Why not use rubber on tractors?* Not a bad question, but in the T's early days rubber tires were fraught with drawbacks; travelers carried as many as four spares on trips of any distance, along with a complete tire repair kit. Nevertheless, the rubber-tractor-tire idea persisted. Homemade "Doodlebug" tractors were often fitted with rubber front tires, and sometimes used truck tires at the rear. Solid (non-inflatable) rubber tires were also used;

some of those were simply solid bands of rubber. Early Fordsons and other standard tractors often labored as industrial "factory mules" on solid rubber tires. Indeed, as early as 1871, a Thompson Steamer that was entered in the California State Fair plowing contest used rubber blocks, like cleats, around the front and rear wheels. But in soft earth, hard rubber couldn't provide the necessary traction.

Orchard growers gave tractor makers something else to think about: Steel wheel lugs damaged tree roots during cultivation. The orchardmen began mounting discarded truck tire casings on their steel wheels in such a way that the

natural strength of the rubber arch supported the weight of the tractor. Several casings could be used side-by-side in order to get sufficient support. These proved so successful that in 1931 B. F. Goodrich brought out a "Zero Pressure Tire."

Tractor maker Allis-Chalmers began experimenting with real pneumatic tires in 1930, and eventually mounted a set of Firestone airplane tires on the rear of an Allis-Chalmers Model U. A-C worked with Firestone to refine these into true low-pressure tractor tires. A-C subsequently announced that for the 1932 model year, rubber tires would be standard equipment on the Model U.

Beginning in the late 1920s, the Lindeman firm of Washington state converted Deere's GP to a crawler for use on the steep slopes of some orchards. Only 24 examples of the GPO ("O" for Orchard) were constructed, and soon after, the Model B was selected for conversion to BO models. Deere would later buy Lindeman and move that company's operations to Iowa for convenience and efficiency.

Nevertheless, a number of farmers were reluctant to make the change. Rubber struck some as alien to a farm environment, and there were dire predictions of polluted land unable to grow crops. Of course, some farmers had made the same prediction about the tractor itself. Fears aside, all major manufacturers offered the rubber tire option by 1934. Just two years later, 31 percent of all tractors were delivered on rubber, and by 1939 almost all were rubber-equipped. The rubber shortage during World War II forced many tractor makers to go back to steel, but they returned to rubber as soon as possible once the war ended.

The popularity of early automobiles like the Model T extended to farmers, who quickly found ways to use the vehicles for more than driving to town. This circa 1920 Pattison Eros was a lightweight car-tractor hybrid that became popular among American, English, and Irish farmers. It and other car-based tractors, also known as "Doodlebugs," put pressure on Deere tractor sales after World War I.

The pulley-mount on the sides of John Deere tractors powered a variety of ancillary equipment. Here, a GP from about 1932 does its duty by spinning an irrigation pump to life. Belts used for the transfer of power were typically made of leather, and when in action they flopped around like a fish out of water. Despite the slack, the system was remarkably successful at delivering horsepower where it was needed.

Economics of the Roaring Twenties

While the Twenties may have roared for some, the decade didn't do much for the American farmer. An agrarian depression that began in 1920 flowed right into the Great Depression of the 1930s. Total U.S. farm income declined from $10 billion in 1919 to around $4 billion in 1921. The root cause for this distress was World War I, which had demanded more farm product. When farmers complied, prices were pushed upward, which encouraged farmers to take advantage of their better position by borrowing to expand their operations. With war's end in 1918, prices fell because of overproduction and farmers were burdened with expensive debt. Further, discharged servicemen returned to farming all over the world, increasing production and driving prices down even further. When farm surpluses were dumped on the world market, buyer reaction to the glut determined the meager prices received by farmers.

Those farmers that could moved quickly to improved labor-saving machinery, such as drills, harvesters, and tractors, a development that was good for Deere and other manufacturers. The number of tractors in use across America rose from 80,000 in 1919 to 800,000 in 1929. This not only drove more U.S. farmworkers into the cities to find employment (increased farm mechanization meant fewer agricultural jobs), but the reduction in the use of horses and mules resulted in a drop in the demand for hay and oats, two staple farm commodities.

When the federal government suspended price guarantees to farmers in 1920, growers with heavy debt loads defaulted, staggering their banks. America's farm depression, coming when the nation's cities were reveling in unprecedented affluence, is one of the unhappiest ironies of the 1920s. This struggling family was photographed in Kennydale, Washington, in 1919.

In the middle of the 1920s, the Deere board began to study the addition of combined harvesters, or "combines," to the product line. The board felt that this move was necessary if Deere was to become more competitive with International Harvester. Combines had been used successfully in the West, but in the Midwest and East, binders and threshing machines remained the norm.

Deere bought several of the better-selling combines and sought to take the best features from each for its own variant. As the new combine developed, blocks of 20 and then 40 were built and sent to the branch houses for testing. A threshing machine was also developed for the traditional farmer not ready to rely on the "newfangled" combine.

In a surprise development, the Holt Manufacturing Company of Stockton, California, which had just merged with the C. L. Best Tractor Company of San Leandro, California, to form Caterpillar, offered to sell its combine line to Deere for $1,256,000. Deere considered the offer but decided that with all they had invested in their own design, the price was too high. The offer was rejected, and Deere continued to refine and sell two versions of its own combine.

This was not the end of it, however. From 1934 to 1935, Caterpillar broached the idea of combining their dealerships with Deere's. "Cat" tractors were of the crawler variety only, so Deere's wheeled Model DI (Industrial) tractor would be a good addition to the Caterpillar product line. Deere, on the other hand, had only wheeled tractors, and under this relationship, dealers could offer Caterpillar crawlers to their customers. Further, Caterpillar had a worldwide dealership organization capable of giving Deere international exposure. In an effort to "sweeten the pot," Cat was also willing to give Deere the rights to its hillside combine, no strings attached. Deere's two combine models were "level-land" types, so Cat's hillside would complete their line. Terms were agreed to, and the arrangement between the two Illinois companies (Cat had moved its headquarters from California to Peoria by this time) proved to be profitable for both, and worked well for many years.

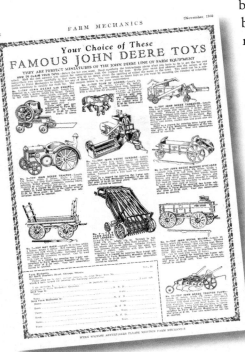

Toys based on Deere tractors and other machines have a venerable history. In 1930 *Farm Mechanics* magazine offered quality tin toys—many with aluminum or nickel pieces—to youngsters who sold *FM* subscriptions.

This ad from the early 1940s shows a Deere Model B pulling a Deere No. 12-A combine, which gave operators broad cuts at heights that could be set as low as 1.5 inches or as high as 40. A Model B summoned sufficient power at the PTO to run the combine with adequate results, but the 12-A did much better work when run off a stationary LUC engine, a variant of the powerplant used in Deere's Model L tractors.

DEERE SURVIVES THE DEPRESSION

With the stock market crash of 1929, investors large and small were, for the most part, wiped out. Immediately and shortly after the crash, though, farmers and agricultural-equipment producers, like Deere, avoided the worst. While crop prices were generally low, most farmers were getting by, and those that could were modernizing. Deere sold about 5,000 Waterloo Boy tractors in 1929, including significant numbers to people engaged in road construction and maintenance.

William Butterworth had been elected to head the U.S. Chamber of Commerce in 1928. The Deere board of directors subsequently gave Butterworth the position of Deere & Company Chairman of the Board. Charles Deere Wiman, the nephew of Charles Deere, was elected company president. The sharply worsen-

Charles Deere Wiman, who served as Deere president from 1928 to 1955, entered the company workforce on the assembly line, where he earned 15 cents an hour.

ing world economic situation was the first test of Wiman's leadership. By 1930 the outlook was dismal in all quarters, and by 1931 the decline finally caught up with the farm-implement industry. Farm prices hit all-time lows over the next three years, and farm bankruptcy and foreclosure sales multiplied.

In 1934, nature worsened the plight of American agriculture. Farmers had unwittingly stripped the Great Plains of the natural grass that not only fed off the rich soil but held it in place. They ended up with barren real estate that had nothing to prevent the soil from blowing away. And blow away is literally what it did. This calamity became known as the Dust Bowl, and was marked by withering drought and blustering winds that sent topsoil whirling into the atmosphere across the Plains, with particular violence in Oklahoma.

Deere & Company experienced operating losses of millions of dollars in the years from 1931 to 1933. The company comptroller blamed the losses on the increased number of tractors and harvesters sold on credit with poor rates of collection. The machines had been purchased in good times, and now farmers were unable to pay for them. Tractors and harvesters were expensive items that required farmers to wring more income from several years of crops, rather than the proceeds of a single year. With diminished crops or none at all across the Plains, payment was impossible. Indeed, many American farms had turned to subsistence farming, growing

(or able to grow) only enough to feed the family.

In a stroke of genius—or perhaps just good luck— Deere & Company made the decision to carry the farmers as long as necessary, and not to foreclose on them or confiscate their equipment. Many farmers, of course, had simply given up and abandoned their land. Unscrupulous predators that bought land for back taxes and forced the farm families off were behind the deaths of many other family farms. But for the most part, farmers were eventually able to pay Deere what they owed. The risk taken by Deere management was to pay off handsomely in the future, in loyalty and respect that eventually made Deere number one in the farm equipment business.

Loose soil sent flying by drought could rise to 20,000 feet and settle as far east as Chicago. The "black blizzards" swept away the futures of thousands of Plains farmers and their families.

Sometime just before the outbreak of World War II, George Farrell's Deere dealership in Prior Lake, Minnesota, mounted this display of a variety of Deere tractor models. Although the Great Depression had tested the company in the 1930s, by 1937 Deere achieved annual sales of $100 million for the first time in its century of existence. The dealerships and tractors, and the green Deere logo, became commonplace in towns throughout the United States.

DEALING WITH THE SOVIETS

Soon after czarist Russia was shaken by the Bolshevik Revolution in 1917, the newly constituted Soviet Union undertook extensive purchases of farm machinery for its collective farms. Deere & Company sent Waterloo Boy tractors and plows, as well as engineers to see that the operators were properly instructed in their use. Payment was 50 percent cash up front and 50 percent on credit, to be paid over the next two harvest seasons. Because Washington didn't yet recognize the Soviet government, Deere assumed great financial risk by entering into this arrangement.

In 1929, the Soviets ordered between 1,500 and 2,200 Model D tractors. Frank Silloway, Deere's sales manager, argued for the deal when the Board was reluctant to trust for repayment in worsening financial times, but

eventually conceded to Silloway's confidence. A few months later, Soviets ordered additional Model Ds, plus a substantial number of plows. Soviet orders on the books amounted to $2 million. Since domestic orders had just about disappeared, the Board had little choice than to set up the sale and hope for eventual repayment.

By 1931, the Soviets had bought $9.8 million worth of Deere products. Soviet farmers and bureaucrats were satisfied that the D worked reliably and efficiently, and with minimum maintenance. Despite the harsh and near-starvation conditions on Soviet collective farms, all contracts were paid on time. The Soviet love of the Deere Model D tractor was a large part of the reason for the timely payment.

The perch for the operator of a prewar Model D hung far off the back section of the frame, but full-coverage fenders kept most of the flying debris a safe distance away. This early example wears the all-steel wheels that were the industry norm before widespread use of rubber tires. Incremental changes were made to the D throughout the model's lifespan, with each alteration adding more versatility to the already-popular model from Waterloo, Iowa.

The Soviet hunger for tractors began with collectivization of farms in the 1920s, and led to inadequate homegrown models, like this one getting the once-over in Uzbekistan around 1930. Deere, however, sold thousands of its Model D to the Soviets, and prospered.

The Model D was ideally suited to farming tasks such as foraging, in which corn or other crops are converted into feed for livestock. This American farmer of the late 1940s pulls a forage harvester attachment designed to pulverize cornstalks and fill a horse-drawn wagon.

John Deere tractors, like this circa 1932 GPWT, could pull two- or three-bottom plows to cultivate the soil. Two unavoidable tasks, lowering the plow into place and then lifting it at the end of each row, were accomplished manually in those days before common farm use of hydraulics, adding potential danger to the equation. Even simple negotiation of a row wasn't without hazards; for instance, not even fenders could have completely protected this farmer from earth that inevitably kicked up around him.

TRACTOR RACES

With low-pressure rubber tires available on its Model U tractor in 1929, Allis-Chalmers organized tractor races at county and state fairs across America to display the ground speed a U could achieve. The specially modified tractors could hit speeds of 60 miles per hour. Famed race car driver Barney Oldfield was one of those put to work piloting the tractors around the dirt horse-racing tracks. He always won—it was written into his contract. Deere didn't sponsor these events, but, as a tractor maker, benefited nonetheless.

JOHN DEERE
MEMORABILIA

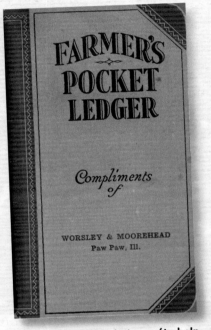

FARMER'S POCKET LEDGER

Compliments of

WORSLEY & MOOREHEAD
Paw Paw, Ill.

Handy, pocket-size ledgers (to help keep track of expenses, animals, equipment maintenance, and many other aspects of farm life) were handed out at Deere branch houses and dealerships. In an unobtrusive touch, Deere didn't tout itself on the cover of this one, which dates from about 1930.

A 1928 edition of *Farm Implement News—Chicago* ran this ad for Deere's 15-27 tractor, better known as the Model D Two-Speed.

This ad from the early 1930s hammers home the promise of increased profits when a Deere spreader is used to properly distribute manure and lime.

Watch Your Crop Yields Climb

Manure is good for your soil. So is lime. They help restore soil fertility—increase crop yields.

Manure, or lime, must be applied properly for best results. This spring, and for years to come, get the full value from your manure and lime by spreading it evenly on your fields with the money-making New John Deere Spreader. Watch your crop yields climb. The increased yields per acre will represent increased profits—that's what you are striving for.

Users say that the New John Deere—the only spreader with the beater on the axle and box-roll turn—is easier to load, easier to pull, does better work and lasts longer.

Spreading lime with the New John Deere Spreader equipped with low-cost lime spreading attachment.

Your John Deere dealer will be glad to show you the many advantages of this popular spreader. See it on your next trip to town. Write to John Deere, Moline, Ill., for free booklet S-14.

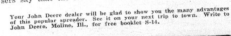

A styled Model B, with jury-rigged spotlight, works the crops after dark.

In this June 1927 farm photo, a Deere Model C with cultivator is staged and ready to go.

Farm snapshot from about 1928 shows a Deere GP Tricycle tractor with front-end harrow setup.

Farmers hate grasshoppers with a passion, so what better novelty postcard image than one of the voracious critters as the "new farm hand"? The card dates to about 1930.

A NEW FARM HAND.

PHOTO BY F.D. CONARD GARDEN CITY, KS.

Deere ads often emphasized the greater efficiencies farmers enjoyed with the company's products.

As Deere accelerated its transition from the "General Purpose" designation, ads and brochures highlighted the new models: A, B, and G. This ad piece is from 1937.

POWER FARMING

JOHN DEERE

WITH GREATER PROFIT —

JOHN DEERE POWER MACHINERY SAVES TIME, WORK and MONEY

JOHN DEERE TRACTORS

GENERAL PURPOSE

NORTH

MODELS "A," "B" and "G"

"BETTING THE FARM" ON MODELS A AND B

In 1929 seven major, longtime implement companies vied for business. They ranked as follows:

- International Harvester   52%
- Deere & Company   21%
- J. I. Case   8%
- Oliver Farm Equipment   8%
- Minneapolis-Moline   4%
- Massey-Harris   4%
- Allis-Chalmers   3%

Interestingly, each one of these companies was engaged in a search for the perfect "all-purpose" tractor, a sort of mechanized Holy Grail that might bring unthreatened industry dominance.

After the tractor price war of 1922, business for the implement makers had been good, although Charles Deere Wiman wasn't assured that Deere & Company would always hold the number-one industry slot. But it wasn't Harvester that was the real threat in the early 1930s but, as we've seen, the economic depression that ravaged American farmlands.

By 1933 overall industrial output across the USA had been cut in half contrasted to pre-1929 levels. General unemployment flirted with the 25 percent mark. Prices, especially farm prices, plummeted. It was in this perilous context that Deere president Charles Wiman made a fateful decision. He instructed Theo Brown, the company's research manager, to aggressively pursue the development of two new general purpose tractors. For Wiman personally, the program allowed him to express his appreciation for machinery and his training as an engineer. But if he was wrong about the eventual success of these new tractors, the company could go bankrupt.

Wiman's plan called for two machines: a two-plow, the Model A; and a one-plow version called the Model B. These were designed as replacements for the disappointing GP and to counter the successful Farmall and Oliver "row-crop" tractors. Although Wiman was forced to eliminate some non-machinery research

> "It would **not be fair** for the farmers to pay for luxuries purchased with **money produced by our implements** before they discharge their **obligations to us.**"
>
> —Kansas City branch manager M. J. Healey, 1931

and to close some experimental farms, he remained firm in his conviction that power farming was the wave of the future.

The Model A came out first, in April 1934, followed by the B almost a year later. The Model B filled a niche not only because of its smaller size but also because it cost less and used less fuel than larger tractors, including Deere's own GP.

JOHN DEERE "AO" ORCHARD TRACTOR
Specially designed for orchardists is this John Deere Orchard Tractor. Seat and platform are low down. Wide, orchard-type fenders prevent damage to trees. Steers easily. Independent differential brakes make possible extremely short turns. Handles the load that ordinarily requires 6 mules. Four speeds forward adapt the tractor to a wide variety of uses. Special citrus fenders can be furnished as shown.

Deere's GP (General Purpose) designation was introduced in 1928, and showed up on A and B Series tractors from 1934 to 1938. For specific farming needs, a variety of front- and rear-axle GP/A combinations could be ordered. AOs like the one seen here (*below and opposite*) did orchard work, and had rear fenders that protected trees; more fulsome fenders were available to citrus farmers. Independent brakes on the rear axle allowed the AO to turn sharply in order to navigate the narrow rows of trees.

The Deere AO, which went on sale in 1935, was configured for use in orchards. It was based on the successful Model A, but with enclosed fenders that protected the low branches of the trees being cultivated. By 1944 the AO carried a 321-cubic-inch, two-cylinder engine, and was capable of 12.3 miles per hour in sixth gear. Reverse allowed a top speed of 4.0 mph. All AO engines ran on gasoline. The AO's vertically sited steering wheel did what it needed to do, but failed to consider the comfort of the driver.

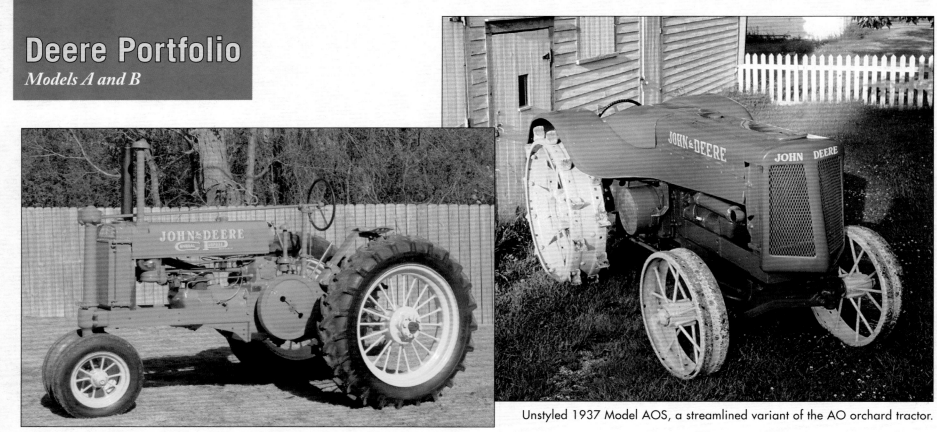

Unstyled 1937 Model AOS, a streamlined variant of the AO orchard tractor.

An unstyled 1935 Model A with "General Purpose" markings.

This Model A Hi-Crop is from 1952, the final year of Model A production.

1945 Model A, with drilled, saddle-style seat.

Right: Deere introduced the Model B in 1934 as a smaller alternative to the A. This is an unstyled 1935 B.

Above: The styled Bs wouldn't appear until 1938, but the cowling of this unstyled 1935 B is as plainly appealing as it is functional.

Here is a 1938 row-crop Model B with the newly styled, rounded front end, partial fenders, a single "tricycle" wheel up front, and metal-only wheels all around.

PTO shaft of the '35 Model B, with implement hitch beneath.

When Brown and his team began work on the new tractors in 1933, general purpose, or "row-crop," tractors, had come into their own. By 1935, however, it was clear that there was a market for row-crop tractors as well as for standard-tread tractors. Therefore, standard tread, or "regular" versions of both the A and B were introduced, as the AR and BR. The company didn't consider these to be "general purpose" tractors because they didn't incorporate the implement lift, adjustable wheel tread, or differential steering brakes. Orchard versions, the AO and BO, were also available. Lindeman Brothers converted some 2,000 BO models to their Lindeman Crawler by adding crawler tracks. Industrial wheel-tractor versions of the A and B were also available, as the AI and BI.

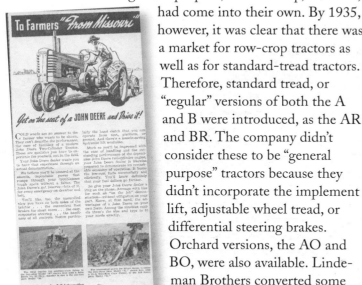

The tricycle configuration was the conventional arrangement for general-purpose machines at the time, with the two front wheels sitting close together. By 1937, variations with only one front wheel, with adjustable wide-fronts, and extra high-clearance (Hi-Crop) versions of all front-end types were made available.

The Deere Model A was introduced first because of the limitations of company manpower and facilities. The Model B was originally available with pneumatic tires. Like the A, it had a four-speed transmission, a power takeoff (PTO) setup, and belt pulley. Its engine, a scaled version of the A's, had enough

Far from crops, this Deere (probably a Model D variant) was pressed into street-washing duty around 1935. Water from the large, towed tank flowed through small-diameter tubes until reaching the spray nozzles up front. Gravity and tube pressure produced adequate coverage and plentiful spray power.

power for two 16-inch plows while the A was capable of pulling four 16s.

What made Models A and B so special? After all, the Farmall had been out for ten years by the time the A was introduced. The answer is that Deere's two new models had remarkable features that were industry firsts: fully adjustable rear-wheel widths on splined axles; and hydraulic implement lifts. These alone were enough to ensure the success of the A and B. In addition, the one-piece rear axle housing provided more crop clearance and allowed a center location for attachment of the drawbar, as well as a center location for the PTO.

In the late 1930s Deere's two-cylinder tractors entered their glory days. This ad from about 1938 emphasizes the ease with which power was transferred from the Model A to trailing implements.

Operation of a typical tractor in sandy soil could be almost impossible, so Deere needed a machine suited to work farms with those conditions. Hence the tracked crawlers, like this 1947 Model BO, which was converted from wheels to treads by the Lindeman Power Equipment Company in Yakima, Washington. (Note the "Lindeman" name on the plate between the drive wheels.) Conversion nearly doubled the cost of the tractor, but the tweaked Deere model was well received by those who needed a broader footprint. When Deere elected to phase out the BO model after World War II, Lindeman anticipated a devastating decline in revenue because of the complexities involved in conversion work performed on tractors other than the B. To preserve Lindeman's expertise, as well as ensure the continuation of reasonably priced conversions, Deere bought Lindeman in 1947 and moved the company to Dubuque, Iowa.

Instrumentation on the standard, wheeled BO was simple, giving the operator oil pressures and other basic information. This panel is from a '47 BO; later models carried instrumentation that was much improved.

On a snowy patch in the late winter of 1948 a Deere Model A (about 3,700 pounds) tangled with a 1940 Chevrolet (about 3,100 pounds). Apparently, nobody was hurt, but the Deere's starboard exhaust stack—as well as the Chevy's front end—took a beating.

In 1936, just two years after the debut of the Model B, world-class designer Henry Dreyfuss was hired to freshen the looks of Deere's workhorse series. The altered sheet metal made an immediate sales difference, as farmers were drawn to the sleeker style of new B models that came out in 1938. This

1941 B has the revised grille and hood that came from the Dreyfuss studio. The '41 Bs also offered a six-speed gearbox for tractors equipped with rubber tires. Steel-wheel variants had to make do with the previous four-speed box.

FRESH STYLING AND ENGINEERING

Production of the Model A and Model B would continue through 1952. Almost 300,000 Model A tractors would be built, and more than 300,000 Model Bs. The tractors were immensely profitable for both Deere and the farmers who bought them. The years brought a continual process of upgrade and improvement. One great line of demarcation occurred in 1938, however, when styling came to the John Deere tractor.

In consumer nations around the world, industrial design had come into its own as a combination of utilitarianism and a new sort of manufacturing aesthetic. By the late 1930s, pleasing proportions were in vogue for products as varied as trucks and telephones, office furniture

and automobiles, tableware and lipsticks. Some designers were achieving prominence and becoming known to the general public. One of these was Henry Dreyfuss of New York City, who was approached by Deere engineers for design help. The engineers' boss, Deere manufacturing chief Charles Stone, had reluctantly allowed his staff to approach Dreyfuss, reasoning that Deere tractors probably did need facelifts.

Dreyfuss struck a deal with Deere and produced a wooden mockup of a restyled Model B within a month. The practical beauty of it startled the Deere engineers. The styling was indeed functional, but amounted to considerably more than just the addition of a radiator grille. The

striking new hood was slimmer, to enhance cultivator visibility, and the new radiator cover protected the cooling system from flying debris. Ergonomics was also considered and greatly improved, with special attention paid to placement of gauges and controls. Dreyfuss helped convince Deere that modern-looking tractors would appeal to young men in farm families, and encourage them to "stay on the farm."

The Deere-Dreyfuss team brought the restyled A and B models out for the 1938 season, a remarkably quick design and manufacturing turnaround. Enthusiastic customer acceptance encouraged Deere to continue its styling improvement program.

Of all the variations seen for the Model B, this one, the BWH, is one of the rarest. Due to the specific nature of its usage as a Hi-Crop machine, only 50 examples were produced. The wide spacing of the front wheels and exceptional clearance made the BWH virtually purpose-built for corn, cotton, and sugarcane. This example is one of the first "styled" tractors to be built by Deere. The sleeker design was penned by Henry Dreyfuss and Associates, and made the Deere line even more appealing.

INTERNATIONAL TAKES NOTICE

Deere didn't pull off its fresh styling in a vacuum, and International Harvester, for one, wasn't caught napping. I-H recognized the trend toward functional and eye-pleasing design, and engaged the services of Raymond Loewy, another noted industrial designer whose shop was already at work on the now-famous "bull's-eye" redesign of the Lucky Strike cigarette pack that was introduced in 1942, and who would later add to his worldwide reputation with the rakish '53 Studebaker Starliner coupe. International Harvester commissioned Loewy to overhaul the entire I-H product line, from the company logo to machine ergonomics. By late 1938, the same year that John Deere brought out "styled" tractors, Harvester introduced the new TD-18 crawler tractor. The rest of the refreshed line followed a year later.

Although Deere was naturally mindful of the new push by International, it allowed the restyled Deere Model A to go on much as before—at least under the skin—for three years. The B, however, picked up an increase in engine displacement, from 149 cubic inches to 175. In 1940, the A got a bump, as well, from 309 to 321 cubes, and a year later the transmissions of both models went from four speeds to six. The transmission change was initially accomplished via a three-speed gearbox and a high-low-range auxiliary gearset. Two shift levers were used. Later, a single-lever arrangement was created.

The Model A finished its career with this engine/transmission combination, but the B enjoyed one more displacement increase in 1947, to 190 cubic inches. Also, both models went from a channel-iron frame to a pressed-steel frame. These 1947 to 1952 Models A and B are known as "late-styled" versions.

With production beginning in 1934, the Model A had already enjoyed a long life when this 1948 example showed up at dealerships. Over the years, revisions and upgrades had happened on a regular basis, with the most obvious being the Dreyfuss-styled redesign of 1938. Operator comfort became more important after the war, so the earlier pan-shaped steel seat was replaced by a padded unit that spared the operator's rump and lower back.

MODELS L AND LA

In 1936, engineers at Deere's Moline Wagon Works had an idea for a tractor to replace a single horse for jobs on very small farms, or for a chore tractor on larger farms. Their goal was a price tag of $500, or less. The Deere board approved the task, with one stipulation: It must run with a two-cylinder engine.

Taking a clean-sheet-of-paper approach, the Moline team settled on a small, conventional running gear with a wide front end, a straight axle, and downward extending kingpins (for crop clearance). This arrangement would eventually replace the row-crop configuration and would be known as the "Utility" configuration. A tubular frame was devised and a two-cylinder Novo engine was mated to a Model A Ford transmission. The steering gear and steering wheel were also Model A Ford. The use of Ford parts greatly speeded the development process of this "Model Y" prototype tractor.

The Model Y was refined into the Model 62, which used a more powerful Hercules two-cylinder engine and a Deere-built transmission and steering mechanism. A few of these were sold to the public. In the 1937–38 model years, a further improvement to the "62" resulted in another designation; the Model L. In late 1938, the L received the Dreyfuss styling touch. Then, starting in 1940, the more powerful and heavier Model LA, with a Deere-built engine and its own serial number series, was introduced.

The L continued, running by now with a Deere-built engine. These were the first Deere tractors with vertical engines, a foot clutch, and a driveshaft. They were fueled by gasoline only. Besides the GP, they were the only Deere tractors to use L-head engines, in which valves are sited on the engine block, next to the pistons, rather than in the cylinder head. The cylinder arrangement that results creates a recognizable upside-down L-shape, hence the name "L-head."

During the mid-1930s, John Deere Wagon Works engineers worked to develop a new, low-cost tractor product. The result, a few dozen units of the promising Model Y, was renamed the Model 62 (*shown here*) in 1937. The rare departure from letter-names for tractor models was reversed a couple of years later when the design was dubbed the Model L. That tractor, which added styling improvements by Henry Dreyfuss, sparked a wartime surge in Deere sales.

Domestic sales of tractors plummeted during the Depression, so Deere and other manufacturers felt pressed to come up with low-priced machines that would prove their worth in the fields. Deere's answer was the Model L, which appeared in 1937, simultaneously with the (1937-only) Model 62. Early L models were unstyled, with an exposed radiator. From 1938 until production ended in 1946, styled versions carried a shapely radiator enclosure and a curved cowl. The L was powered by a vertical, two-cylinder motor fed by gasoline only. A foot-operated clutch allowed the operator to choose among three forward gears. In all, 13,365 copies of the L were produced, selling for between $475 and $517.

Deere Portfolio
Models L and LA

Left: The first Model L tractors were unstyled, and closely resembled the 62.

Below: The Model 62 was built during 1937. It was the second of a follow-on series of three small row-crop tractors that began with the Model Y in 1936, and culminated in 1937 with the Model L. All Model 62s ran with a 10.4-horse Hercules engine. This one sports full fenders.

A styled 1941 Model L, with single-bottom plow.

Henry Dreyfuss designed the styled L models to look markedly different from other, larger Deere tractors. This one, with optional lamps, dates from about 1940.

The Model LA was introduced for 1941, and was built simultaneously with the L until both models departed in 1946. The saddle of this '46 LA has been carefully fitted with padding.

Dad positions himself in front of the rear wheel as Junior fires up the unstyled Deere General Purpose in about 1938.

A restrained yet clearly proud Deere bowling team, probably snapped in the early 1930s. Clockwise, from top left, they are Mary, Clara, Alice, Nell, and Mary.

Good times with Dad, son, stripes, and a Model B.

Wildorado, Texas, sits on Route 40 about 15 miles west of Amarillo. This snapshot was taken there, probably by a proud husband and father, in 1939.

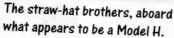

From 1944, a family photo of an early Model A (with steel, rather than padded, saddle), fitted with rear- and mid-mounted cultivators.

The straw-hat brothers, aboard what appears to be a Model H.

Deere's centennial, which arrived in 1937, was one focus of this thick and colorful promotional booklet.

LEADERS

In MODERN DESIGN... PROVED PERFORMANCE

GET acquainted with a *really modern tractor*—try a John Deere! Up on that deep-cushion seat, you'll relax in comfort. There's a perfect view ahead and to either side. Every control is right where you'd naturally want it. A roomy platform lets you stand up as easily as rising from a chair.

In the field, under a heavy load, you'll be amazed at the power and smoothness of the John Deere Cyclonic-Fuel-Intake Engine. Touch the hydraulic control and watch how effortlessly Powr-Trol raises, lowers, and adjusts your integral or drawn implement. Find a rough field and experience "Knee-Action" ease of handling and comfort as those Roll-O-Matic Front Wheels "walk" right over obstructions, smooth out the bumps. Try the multi-speed transmission that provides a speed for every job. Make any other comparisons you care to make and you'll agree that a John Deere Tractor has no equal for modern, functional design.

But that's just half the story. An exclusive, time-tested John Deere principle—*two-cylinder design*—provides a degree of simplicity and strength unequalled in other tractors. There are less than half the number of moving parts in the engine alone and, size for size, each part is correspondingly larger, heavier. This means less wear, less trouble, fewer replacements—greater dependability, greater economy, longer life.

It is this balanced combination of *modern design* and *proved performance* that has earned for John Deere Tractors the respect of farmers everywhere—that will make you want a John Deere once you know all the facts. See your John Deere dealer or write John Deere, Moline, Illinois, for further information.

JOHN DEERE *two-cylinder* **TRACTORS**

A 1949 Deere ad celebrates Deere innovations in design and technology, with special attention given to the company's long-standing commitment to two-cylinder engines.

Farmers who had Deere portable mills saved time and money by converting their silage into feed for cows and hogs. This ad, from about 1940, declared that Deere mills are "mighty economical."

Shrink FEED-MAKING COSTS

ONE sure way to increase milk and meat producing profits is to lower feeding costs. You can make your home-grown feeds go farther and cost less by processing them yourself with a big-capacity John Deere Hammer or Roughage Mill. Because of its low initial costs . . . low power requirements . . . and low upkeep costs, a John Deere Mill is mighty economical to own—it's a mill that will soon pay for itself in the better feeds it makes and in the money it saves you.

See the full line of profit-making John Deere Mills at your John Deere dealer's. There's a size and type to match your needs. Mail coupon below for free folders.

JOHN DEERE HAMMER and FEED MILLS

John Deere, Moline, Ill., Dept. F-25.
Please send me Free Folder I checked below:
☐ 10-inch Roughage Mill
☐ 14-inch Roughage Mill
☐ 10- and 14-inch Hammer Mills

Name ....
Town ....
State .... R. F. D. ....

FURROW
SPECIAL EDITION DEDICATED TO THE AMERICAN FARMER

AMERICA'S No. 1 WAR WORKER
WHAT *He* Does... HOW *He* Does It... and WHY

This 1943 issue of Deere's *Furrow* magazine celebrates the priceless wartime contributions of America's farmers.

MODERN HARVESTING TIME—THE GOLDEN TIME
"AS YE SOW, SO SHALL YE REAP"

"Modern Harvesting Time—The Golden Time." A styled Model B with Deere harvester, from about 1940.

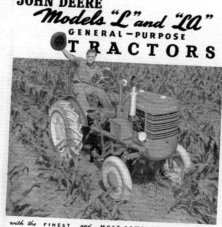

JOHN DEERE
Models "L" and "LA"
GENERAL-PURPOSE TRACTORS

with the FINEST *and* MOST COMPLETE LINE *of* WORKING EQUIPMENT *in the* SMALL TRACTOR FIELD

The babies of the Deere tractor line from 1937 to '46 were the L and LA. The LA seen here rides on sturdy, cast rear wheels.

German engineer Dr. Rudolf Christian Karl Diesel invented the oil-fueled engine that bears his name. In 1892, he patented an internal combustion powerplant that auto-ignited fuel by using the heat of compression. Diesel continued to refine his engine, and made it commercially viable by 1897. It was so large and heavy, however, that it was considered only for ships and stationary power plants. Over time, of course, diesels of manageable sizes became possible.

The main advantage offered by the diesel is its frugal fuel consumption due to thermal and volumetric efficiency that surpasses that of gasoline engines. The fuel oil burned by diesels has more heat value (BTU) than gasoline. Another advantage is that, while kerosene and gasoline engines run fuel-rich for cooling purposes, diesels run lean. Further, efficient fuel injectors, rather than a throttle, control engine speed, bringing consistency to any rpm.

Deere's first diesel was installed on the 1949 Model R tractor.

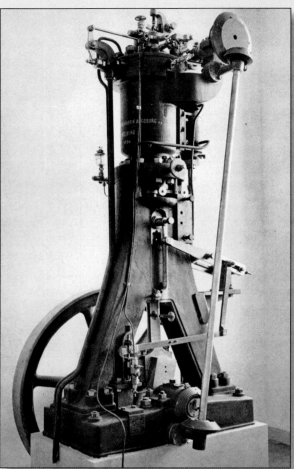

Because Rudolf Diesel (*right*) licensed as well as patented his innovative engine (*above*), he became a millionaire. His first engine was powered by peanut oil, and proved that fuel could be ignited without a spark. Instead, combustion occurred inside a cylinder. Diesel remained a vocal proponent of vegetable oils as fuel.

MODELS G AND H

As the 1930s ended, progressive farmers generally had outgrown their horses. Now they were in the market for tractors that would handle larger implements, move them faster, and further multiply the efforts of their manpower. The 1937 arrival of the big John Deere Model G was appreciated by the larger-acreage farmer who needed at least some row-crop capabilities.

Farmers also saw the need for a small tractor, to replace just one or two horses for small jobs. The Model H of 1939 appealed to small-acreage farmers; it was the power equivalent of the original Model B, which had grown to be the power equivalent of the original A. The G and the H were available as row-crops only.

The G was one of the last existing Deere lines to receive Dreyfuss styling, in 1942. At the same time, its transmission was upgraded to six speeds and the tractor was re-identified as the GM ("M" for modernized) in order to get a price increase past the War Price Board. After the end of World War II, the GM name reverted to G, or to New G.

The unstyled and GM versions were available only as dual-front tricycle types. Postwar versions were also available in single front wheel, wide front, and Hi-Crop arrangements.

The Model G came on line in 1937, and received Dreyfuss styling for 1942. The G was Deere's attempt to build a tractor as powerful as the D, but which carried fewer pounds. The new tractor succeeded by creating nearly identical horsepower while saving half a ton in weight. An all-fuel machine, the G ran economically on distillates, but produced more power with gasoline. The G was such a popular model that it was produced until 1953, long after the A and B had been culled from the Deere lineup.

The Model H was introduced in 1939, and was aimed squarely at farmers who worked spreads of fewer than 80 acres. The H was a capable tractor with plenty of flexibility. It was available in five configurations ranging in price from $595 to $650. Rated at a maximum belt horsepower of 14.2, the H was well suited for those smaller farms. Deere built the H between 1939 and 1947, with total production approaching 59,000 units.

HYDRAULICS, AND A COUP FOR FORD

Probably the greatest improvement in tractor utility was hydraulic power. The first tractor hydraulics appeared in 1934 on the John Deere A, as an implement lift. A fairly low-pressure setup, it was enclosed in the tractor's final-drive housing to keep it out of the dirt. Other manufacturers followed with similar systems, culminating in the famous Ford-Ferguson three-point hitch of 1939.

Ferguson's patented system had unique feedback control. Inappropriate movement of the control lever created an error signal in the valve pack; in response, the hydraulics moved the implement to correct the error. The secret of the system was that implement overload could also create the error signal. Thus, if the plow hit harder soil, it would raise automatically until the draft load was the same as had been previously

set by the operator. This raising and lowering to maintain the set draft load happened instantly, without operator input. It made tractor plowing a job even the inexperienced could do. Henry Ford, in fact, had an eight-year-old boy demonstrate plowing with the new Fordson 9N at the press introduction in 1939.

The Ford-Ferguson was possibly the most significant tractor of the 20th

Because the 1942 Fordson 2N was an all-new tractor, it didn't fall under wartime price-freeze guidelines for existing tractors that had been established by Washington. Henry Ford himself lobbied successfully for scarce materials, ensuring that the 2N would be marketed with a battery and rubber tires—big advantages in a much-changed American tractor marketplace. Naturally, none of this was received as good news by Deere. The 2N remained in production until 1947.

Henry Ford (*right*) and his partner, safety-hitch inventor Harry Ferguson, inspected this Fordson tractor equipped with Ferguson hardware in about 1939. Although Deere and others would eventually come up with technology that superseded the Ferguson hitch, when Harry's invention was new, it was a revelation across the tractor industry.

century. It initiated the change from the row-crop configuration to the now-standard utility configuration, and it introduced the load-compensating three-point hitch. Although small (2,500 pounds and with a 120-cubic-inch engine), it could plow 12 acres a day with a 2–14 plow. This was a rate that Deere couldn't equal with anything less than its Model G.

More than 840,000 Ford-Fergusons were sold between 1939 and 1951, at an initial price of less than $600, rising over the years to about $1,200. At the peak of production, 9,000 Ford-Fergusons were delivered every month.

The massive John Deere Harvester Works plant rose on a 43-acre site in Moline in 1912, the same year the modern Deere & Company was created. The back of this circa-1939 postcard reads in part, "This plant produces harvesting machines for practically every crop. John Deere grain binders, rice binders, corn binders, corn pickers, combines, threshers, mowers, self-dump rakes, field ensilage cutters, and mower knife grinders are made here."

The Union Malleable Iron Works, shown in this 1939 postcard rendering, turned out malleable iron materials used in other John Deere factories for more than 75 years. Founded in 1872 as a separate corporation, with Charles Deere as one of its original principals, the iron works was purchased by Deere & Company in the merger of 1911. The original plant was demolished and replaced by a new foundry in 1968.

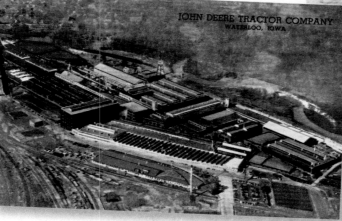

The John Deere Spreader Works brought under one roof the capabilities of two Deere acquisitions. Syracuse, New York-based Kemp & Burpee Manufacturing Co. was purchased in 1911 on the strength of its commercial success with the first mechanized manure spreader (a big hit with farmers). Deere improved on the design and chose to manufacture the new spreaders at the Marseilles manufacturing plant in Moline, which Deere had purchased in 1908. The product was so well received that in its first year the line turned out 23,000 spreaders. The facility was later renamed John Deere Spreader Works.

It's only about 150 miles from Moline, Illinois, to Waterloo, Iowa, but Deere's 1918 purchase of the Iowa tractor company was a quantum leap for an enterprise that had been built on plows, planters, and wagons. The John Deere Tractor Company facility in Waterloo depicted in this 1939 postcard was then the largest tractor factory in the world.

WORLD WAR II AND POSTWAR EXPANSION

During World War II, Deere & Company launched itself into the defense effort, producing a wide variety of items, from howitzer ammunition to laundry units, and from the MG-1 military tractor to tail-wheel assemblies for P-47 Thunderbolt fighters.

Charles Wiman resigned the presidency for the duration and, as a colonel in the U.S. Army, managed Deere's tank and combat vehicle division. Burton Peek was elected as Wiman's replacement.

To ensure a continuing supply of Deere products for the military, the Army offered to share the cost of a new plant. The chosen site, Dubuque, Iowa—though outside either the Moline or Waterloo labor markets—was close enough for ready control from Moline. Location on the Mississippi River also was a factor that worked in Dubuque's favor. War materiel flowed out of the new plant, which was built with the understanding that, after the war, Deere would retain the facility.

In 1944, Charles Wiman once again took the helm at Deere. To counter Ford's inroads into the small-farm marketplace, Wiman oversaw the new John Deere Model M, a replacement for the L, LA, and the H. The M was billed as a general-purpose utility tractor. It had a gasoline-only, vertical, relatively high-speed (1650 rpm), two-cylinder

The Model M was Deere's first postwar design and would replace the L, LH, BR, BO, and H models in one fell swoop. MT models like the 1950 example seen here were most often in the three-wheeled "tricycle" configuration. The MA and MT had a telescoping steering column that permitted the driver to stand as he worked. Not the safest way to go about one's business, but it did allow a much more expansive view of the field. A new creature comfort was a backrest that could be inflated to the operator's desire with an internal air bladder. The M series was built between 1947 and 1952, and did well against competing tractors from Case, Ford, Allis, and Farmall.

Nearly 12,500 examples of the Deere Model LA were produced from 1941 to 1946. The tractor, which sported smart Deco styling by Henry Dreyfuss, was well suited to modest acreages planted with beans, strawberries, and similar crops. It also was an effective hay cutter and general-maintenance machine. The two-cylinder LA engine produced 14 horse-power, a nearly 40-percent increase over the 10.4-horse Model L motor, and nearly as much as the output of the larger Deere Model H. In 1946 a buyer could bring home an LA for $650.

engine—a departure from the customary horizontal transverse engine. It also had the convenient "Touch-O-Matic" rear-implement hydraulic lift. Although the lift was similar in function to the 3-point hitch, it didn't incorporate Draft-Control, as the situation vis-à-vis Ferguson's patents was unclear.

Despite that ambiguity, the M was configured much like the Ford-Ferguson. The M's vertical inline engine necessitated a driveshaft to bring the power back to the transmission. The clutch was foot-operated. These features had been pioneered on Deere tractors by the diminutive garden-type Models L and

LA, which also had vertical, inline two-cylinder engines. The Model M and its successors were built in Deere's Dubuque plant.

The M's configuration didn't satisfy all of Deere's customers, however, so in traditional Deere fashion another machine, the MT, was added to the line in 1949. The MT was essentially the same tractor as the M, but could be equipped with an adjustable wide-front, dual-tricycle front or single front wheel. Dual Touch-O-Matic was an added option that allowed independent control of right- and left-side implements.

Dubuque, then, became the home of

Deere tractors with driveshafts, such as the Model M and the related machines that followed. Note that the other Deere tractors, made at Waterloo, had transverse engines that were geared directly into the transmissions. This production arrangement was continued until 1960, when Deere discontinued its venerable two-cylinder engine.

Models A, B, D, G, and M were replaced with equivalent tractors with number identifiers. The D was initially replaced by the Model R diesel, which later became the Model 80 in the numbered series.

THE INDUSTRIAL EQUIPMENT DIVISION

The Yakima, Washington-based Lindeman Power Equipment Company enjoyed wartime success as a maker of crawler-style tracks and carriages that it fitted to Deere tractors. Lindeman was essentially offering conversion kits to the numerous farmers who wanted a crawler tractor, so Deere did an end run when it purchased Lindeman in 1947, to make crawlers of its own. Generally, these were fitted with 'dozer blades for construction work. (This move more or less ended the 10-year-old relationship with Caterpillar, by which Cat sold Deere tractors for industrial uses, thus avoiding the expense of developing its own small-tractor line. The arrangement had been reciprocated, as Deere product gained access to Cat dealerships.)

The takeover of Lindeman allowed Deere's entrée into log skidders and other specialized forestry equipment. Later, road graders, excavators, and scrapers were added to the line.

In 1956 Deere & Company went after the industrial equipment market by offering "yellow" versions of its regular farm tractors for highway mowing and for earthmoving.

Deere's popular and versatile Model B was a natural for conversion by crawler specialist Lindeman. Deere bought that company in 1947, at just about the same time that this converted '45 BO became available. This example, fitted with a 'dozer-style blade, was built for reasonably heavy work. The crawler's cockpit (*left*) was a rat's nest of control levers, each contoured and canted so that the operation of one didn't interfere with any other. The BO and similar GPO crawlers were primarily used in orchard settings, or where sandy soil mandated good stability and traction.

DIESEL TRACTORS

The first diesel tractor was the Caterpillar Diesel Sixty of 1931. It was followed in 1934 by the McCormick-Deering WD-40. The WD never gained the popularity of the 1941 I-H Farmall MD, which revolutionized farm power and normalized the use of the diesel engine. Sales of the Farmall MD averaged 22,000 units per year, even though the MD cost 50 percent more than a gasoline version. Fuel consumption was about one-third to one-half that of the gasoline version.

Deere's first diesel tractor was the Model R, introduced in 1949. It was based on developments that Deere had undertaken as early as 1940, with eight experimental MX models that were tested and then redesigned to address shortcomings.

It's interesting to note that the final engine design for the Model R's two-cylinder diesel employed the same bore, stroke, and valve sizes as the six-cylinder Caterpillar D8. Some historians believe that informal cooperation between the two Illinois companies continued after Deere's 1945–47 move into industrial equipment.

The 1949 Model R was Deere's first foray into the world of diesel power. The pluses of diesel economy and torque far outweighed the negative issues of weight and hard starting in cold climates. A small two-cylinder, gas-powered "pony" engine was used to bring the 416-cubic-inch diesel motor to life. Although the R weighed in at nearly four tons, it could nevertheless reach 11.5 mph when rolling on rubber tires in fifth gear. Steel-wheel R models had a factory-blocked fifth gear, to limit top speed to 5.3 mph.

Deere first offered diesel power in 1949, on the Model R. Because diesels were famously reluctant to start in cold weather, Deere's unit was preheated by a gas-powered "pony motor" (which was itself electrically started). Dependable starts are central to this 1955–56 ad for Deere's Model 80 tractors.

The Model 4010 diesel (*left and above*) hit dealerships in 1960, after years of development. Deere initially experimented with V-oriented diesels of four, six, and eight cylinders, but finally settled on a straight six after finding that dispersion of gases in the V-blocks was inefficient. Comfortable, adjustable seating, improved pedal ergonomics, and an 8-speed Synchro-Range transmission helped ensure the 4010's success in the marketplace. The tractors were built at Waterloo and in Mexico.

Deere produced the Model 60 from 1952 to 1956, eventually offering it in four iterations: standard, Hi-Crop, orchard, and row-crop.

Like the 60, the Model 50 of 1952–56 had Deere "Power Steering" (labeled at the side of the radiator, below the model number).

In 1959–60, a new Model 630 cost the owner $3,300. The tractor had six forward speeds and produced 48.7 horsepower at the PTO. This row-crop variant has the familiar 30 Series "Oval Tone" muffler.

JOHN DEERE
MEMORABILIA

GETTING YOUR HOUSE IN ORDER

Practical Suggestions
FOR JOHN DEERE DEALERS

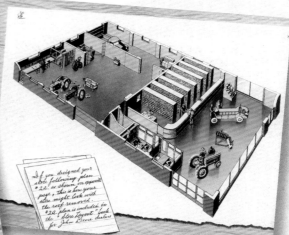

The 420 line arrived in 1955 with an all-green color scheme. By June 1956 the tractors had adopted yellow panels that ran along both sides of the hood and at the sides of the radiator.

In about 1950, Deere provided its dealers with "Practical Suggestions" for modernizing their franchises, and making them more efficient and profitable.

It's POWERFUL Good News!

NOW **NEW POWER...**

NEW 2-3 Plow Tractor Performance That Will Open Your Eyes!

THE NEW JOHN DEERE **420** FLEET

A popular heavy-cast toy of the Model 730 of 1959–60.

CANADA PAINT

TRACTOR AND IMPLEMENT ENAMEL

MEDIUM YELLOW (CATERPILLAR)

ORANGE (ALLIS CHALMERS)

BRIGHT RED (MASSEY-FERGUSON)

DEEP RED (INTERNATIONAL HARVESTER)

BRIGHT GREEN (JOHN DEERE)

ALSO BLACK

DUE TO IMPLEMENT MANUFACTURERS CHANGES IN COLOR STANDARDS THESE COLORS ARE APPROXIMATES ONLY.

This chart created by Canada Paint in about 1950 shows enamel colors suitable for products made by (top to bottom) Caterpillar, Allis-Chalmers, Massey Ferguson, International Harvester, and John Deere.

On July 26, 1949, John Deere board members "and Their Ladies" attended a gala dinner at San Francisco's Fairmont Hotel. Menu items included crab legs and roast larded filet of beef. Among the entertainers were an accordionist, a Hawaiian comedienne, and Chinese acrobats.

A corn picker and a tractor become fast friends in this storybook published by Deere in 1959. Official facsimile editions can be purchased today.

With a Deere Model A as the centerpiece, Wisconsin farmer William Renk (*standing, left*) speaks with his sons and grandson in May 1947. The Renks' prize-winning sheep were known across America for their quality bloodlines.

In 1958, visitors to Deere's Waterloo Tractor Works were presented with this souvenir booklet.

Deere Portfolio
Diesels

The R's diesel displaced 416 cubic inches and produced 51 horsepower.

The Deere Age of Diesel was inaugurated in 1949 with the Model R (though Deere had explored diesel development for more than 10 years before that). This Model R dates from 1951.

Above: The 4020 diesel was Deere's update of the popular 4010, which was among the first of the New Generation tractors of 1960. This 4020 is a Hi-Crop variant, which marks it as about a 1965 model.

Above right: Model 820s from early in the 1958 run had all-green dashes; fascias of later '58s were black. Power steering was standard.

Right: Its graceful fenders aside, the 1958 Model 820 was a tough puller with a 472-cubic-inch diesel that needed a small, gas-fired "pony engine" to get started.

WILLIAM HEWITT TAKES THE REINS

In 1954, Charles Wiman learned that he had contracted a terminal illness and had only months to live. His last major responsibility, after 26 years at the helm of Deere & Company, was to oversee the selection of his replacement. The company's predisposition to Deere family members was uppermost in Wiman's mind, but lineal descent wasn't considered sacrosanct by the Deere board. Nevertheless, William Hewitt, a board member and Wiman's son-in-law, was selected to be the next president. Wiman died on May 12, 1955, and 12 days later Hewitt was installed as president.

Hewitt's qualifications for the job amounted to considerably more than the family connection. William earned an economics degree at the University of California at Berkeley, and had had a distinguished naval tour during World War II, rising to the rank of lieutenant commander. Although young at 40, his experience in the tractor business included stints with the Pacific Tractor & Implement Company, Ford-Ferguson, and, beginning in 1948, as a Deere territory manager in California.

Hewitt was sensitive about his family ties, and made a special point to prove himself a competent and cool leader. As president and, later, as company chairman, he brought Deere through numberless changes in the postwar period, and all the way into the 1980s.

During Hewitt's tenure, Deere's annual sales surged from $300 million to more than $5 billion. Much of the additional revenue was found abroad, where Hewitt expanded aggressively. Hewitt also perceived the inexorable shift from family farms to large-scale agribusiness, and made sure Deere was well positioned to service that trend.

Unique among his successes is the Modernist Deere headquarters, which was designed by architect Eero Saarinen under Hewitt's general direction. At once avant-garde and dignified, the structure sits among rolling hills, exemplifying Hewitt's commitment to Deere's legacy, and to the future.

As John Deere Day films became popular with small-town audiences, the company invested in Hollywood talent to up the ante. The prolific character actor Walter Baldwin (*right*) played the part of a farmer in this 1954 Deere light comedy. The same year, Baldwin appeared with Audie Murphy in *Destry,* and in the Jerry Lewis and Dean Martin romp, *Living It Up.*

Long before the Madison Avenue "Mad Men" era, John Deere marketers used a variety of creative, sometimes wacky, tactics to keep the Deere name in the minds of Americans. After World War II, dealerships often sponsored John Deere Day community events, replete with imprinted giveaways and often centered around a self-promoting Deere movie, such as this 1954 production filmed by Reid H. Ray Film Industries. It got better: One tractor-line launch in 1960 included fireworks, an ice skating revue, and a diamond-studded Deere tractor on the floor of the upscale Nieman Marcus store in Dallas, Texas.

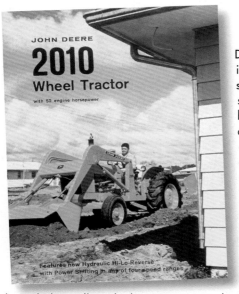

Deere is a longtime player in the industrial tractor segment, which produces sturdy machines needed by road crews, developers, and other demanding users. Some of the Deere industrials shared model numbers with their farming brethren, like this 1960 Model 2010, but their distinctive yellow-orange color set them apart. Four speed ranges and the ability to power shift through them all made the 2010 easy to handle in a variety of tasks. Only seven wheeled 2010 industrials, all gas-powered, were built for 1960, but 403 gas and 291 diesel examples rolled off the line for 1961. Numbers grew nicely throughout the tractor's 1960–65 production run, with 1,207 diesels produced for 1965, with 555 gas models following behind.

The 430C was another Deere industrial based on a wheeled tractor. As with this 1959 example, the 430C was available in a four- or more popular five-roller model. It could also be equipped with several implements. A front-end loading shovel and three-point lift were two of the more common add-ons. A perforated grille guard was often added to protect the radiator from being pierced by debris—a disaster that would put a quick stop to any operation.

Into every life some rain must fall and in John Deere's world the 8010 was a storm. A massive 20-ton machine designed to be the workhorse on any large farm, the 8010 ran with nine forward gears and was articulated for concise handling. Unfortunately, a problem with the gearbox was figuratively bigger than the tractor itself, and all but one of the 8010 models were returned to Deere. With fixes made, the new 8020 returned to the fray and functioned well. It was reduced to eight speeds but the leviathan's dimensions remained the same as its predecessor's: 120-inch wheelbase, 235 inches in length, and 96 inches wide.

Industry Leadership

"When [Deere executives] were tempted to be satisfied with second best, he showed them the greatness that was hidden in them."

*—Management analyst James O'Toole,
on Deere president William Hewitt*

Deere president William Hewitt, who assumed that post in 1955 and built up the company's international presence, is particularly well regarded for shepherding Deere's trend-setting "New Generation" tractor line to market in 1960.

THE COMPETITION STUMBLES

In 1951, International Harvester's Board of Directors repudiated Fowler McCormick, the last of the McCormick family managers, and replaced him with his rival, John McCaffrey, as chairman and CEO. McCaffrey instituted a less deliberate style of management that emphasized engineering more than testing. He also pushed the I-H construction-equipment product line—at the expense, some feel, of the farm line.

By the late 1950s, quality problems became chronic at I-H. When sales fell, factory layoffs and diminishing employee morale followed. Competitors gained ground, though I-H retained the number-one position in agricultural equipment until 1963. At that time, the lead was yielded to John Deere.

William Hewitt, who became president of Deere in 1955, took special

notice shortly thereafter of a piece of I-H dealer propaganda that came across his desk. The motto on the bottom of the page was, "Not Content to Be Runner-Up." This got Hewitt to thinking: Was Deere content to be number two? He concluded that if they were, he would change that!

Nevertheless, the eight-year period from 1954 to 1962 saw some remarkably beautiful and highly competent tractors emerge from the I-H shops. Harvester was riding a wave of customer goodwill when this period opened. In 1955, in fact, the 3-millionth Harvester tractor came off the assembly line.

But things were happening too fast for Harvester. The Model 70 was Deere's top-end row-crop machine and produced more than 50 horsepower when it was introduced in 1953. (I-H's Farmall Super M, introduced the year before, had only 44 horsepower.) Besides, the entire

> **"Not Content** To Be Runner-Up."
>
> *—International Harvester dealer slogan, c. 1955*

After a brief yet excruciating delay, Deere's New Generation tractors made their way onto the market in 1960. The new 3010 was a four-cylinder version of the larger 4010, but delivered on the promise of being a wholly new breed. A new, triple-circuit hydraulic system corrected a long-standing drawback by which all components had had to share the same hydraulic line, which greatly reduced efficiency. The new setup allowed the 3010 to have separate circuits for power steering and power brakes, and a third for the three-point hitch or any other remote implement. When fitted with the LPG (liquid petroleum gas) propane option, the 3010 was the only Deere model that concealed the fuel storage tank beneath the cowl; earlier Deeres made do with a bulbous tank located outside the lines of the long hood.

Various iterations of the Model D ran on gas, LPG (liquid petroleum gas), diesel, and all-fuel. Whatever the engine, all were two-cylinders. Though introduced in 1953, the popular 70 was produced through 1956, and was good enough to keep customers' minds off the delay of the introduction of Deere's New Generation. The 70 had better hydraulics than that of its predecessor, the Model G, as well as a 12-volt electrical system and power steering.

Deere line could now be equipped with the "Power-Trol" hydraulic pump, Deere's answer to Ford-Ferguson's weight-transferring three-point hitch. Harvester's marketing people absorbed all these developments and knew something had to be done just to stay in the game.

Thus, I-H introduced its numbered series in mid-1954, giving the impression that the company was offering a whole new lineup. Not much that was new had been done technically, but there was more power for the larger Farmalls. I-H, however, could not bring itself to copy the Ford-Ferguson three-point hitch, because they had always treated it with disdain. Instead, Harvester engineers came up with a two-point "Fast-Hitch" hydraulic implement lift, which wasn't compatible with three-point implements that by now were the industry standard. It wasn't until 1961 that Harvester offered a three-point option.

In the meantime, I-H had another problem. Because of the increased power of the largest Farmall, the Model 450, drive differentials began to fail. A failed differential meant the affected tractor couldn't move. An inch. Failures quickly became epidemic and farmers who had purchased the Harvester product were angry. Tractors were breaking in large numbers right in the middle of plowing season, and it would be a while before a fix could be devised and produced. To its credit, International Harvester did its best to accommodate customers with loaner tractors and free repairs, but the damage to the company's reputation was done.

By the late 1960s, Harvester was delivering some thoroughly modern and entirely redesigned tractors. Big four-wheel drives were part of the 1966 lineup. In 1980, the articulated, Model 3788 was introduced. It was possibly the most advanced four-wheel-drive tractor made. The "Control-Center" cab was behind the articulation point with the engine and transmission in front. Harvester never recovered their customer goodwill, however, and their market share continued to go downhill.

The massive, articulated Model 8630—one of Deere's Generation II models with fresh sheet metal—was introduced at an industrial show in Saarbrucken, Germany, in the fall of 1972. This 24,200-pound, four-wheel-drive tractor entered regular production in 1975, and came with Deere's brand-new Sound-Gard cabin as a standard feature. The cabin kept sound levels impressively low, even when the 8630 was grunting at its rated 2100 rpm. The 16-speed Quad-Range gearbox was another first.

JOHN DEERE
MEMORABILIA

JOHN DEERE
Manure-Handling
EQUIPMENT

TWO Powerful Loaders
FOUR High-Speed Spreaders
TO MATCH YOUR EXACT NEEDS

Handle and spread your farm's manure properly, and you have
a big leg up on proper cultivation of your soil. This ad appeared
around 1957.

Even as Deere made enormous strides
in the 1960s, toys, like this heavy-cast
replica of an unstyled Model A or B of
the Depression years, happily recalled
the company's past.

Development and
marketing efforts
undertaken in
1965 paid off
for Deere in
1966, when the
company's total
sales blew past
$1 billion for the
first time.

Modern Farming: 1965

By 1963, farms were growing larger, and needed greater efficiencies. The 4020 answered that need, with bumped-up horsepower (91 for the diesel version), and a sophisticated hitch design that allowed hard pulling and quick switch-outs of implements.

John Deere double-up power!

Put John Deere power in front of a minimum-tillage hookup and you can cut production costs as much as $5 an acre. Makes good farming sense—put more of a load on tractors . . . double-up on tools . . . get several operations done once-over.

John Deere saw the handwriting on the wall in the need for bigger tractors and put out models ideally suited to double-up operations. The 65 h.p. "3020"* and 91 h.p. "4020"* deliver power for multiple-unit hookups not only on the drawbar but through the PTO and hydraulic systems as well.

Hand-in-glove with the development of these tractors, John Deere engineered equipment-ganging devices to make big tractor power pay off. Minimum tillage hitches. Squadron hitches that link together up to 40 feet of drills. Tool carriers brawny enough to handle several types of implement attachments simultaneously.

Your John Deere dealer wants you to take a crack at farming with double-up power. He'll supply the tractor and implements if you'll supply a few hours of your time. When the trial is over, he'll make it easy to shake hands on a tractor and implement deal with good John Deere Credit Plan terms.
*Diesel model

JOHN DEERE
Moline, Illinois

This rather sophisticated JD pencil could have been used to make notations in a farm's ledger book.

When sewing a John Deere slipcover or pillowcase, what better helper than a John Deere thimble?

Deere's well-received Power Shift setup "tuned" gear selection to soil conditions "with a quick, almost effortless move of the single control lever." This advert is from about 1967.

What's more important than gallons per hour?

The fuel consumption difference between most modern tractors can almost be measured in eyedroppers per hour. The really important difference shows up in working speeds . . . and that's where John Deere Tractors shine! More usable horsepower at faster ground travel speeds . . . that's the hour-saving advantage of John Deere ownership. Apply the most accurate yardstick of tractor performance . . . acres of work completed per day, not inches of fuel left in the tank . . . and you'll always come out ahead with a John Deere.

See how one measures up on your farm . . . no obligation . . . that's your dealer's invitation. Take him up on it soon . . . and, while you're about it, get the low-down on his income-stretching tractor financing plan.

Acres per Hour!

JOHN DEERE
Moline, Illinois

An ad from about 1966 emphasizes the work efficiency of Deere tractors. "More usable horsepower at faster ground travel speeds" was an advantage pointed out in the ad's copy.

Can tractor shifting be this easy?

**Yes...
with John Deere
Power Shift**

Change gears as quickly and easily as you did between radio programs. "Tune" speed selection to soil condition with a quick, almost effortless move of the single control lever. This single-handed freedom of selection is yours with the new Power Shift Transmission of a John Deere 65 h.p. or "3020" or 91 h.p. "4020" Tractor.

It doesn't matter "where" you want to go—up or down a gear . . . forward or reverse . . . progressively or skipping gears. You name it; Power Shift delivers . . . all without clutching, all without interruption of power to the drive wheels. You can bring any of eight forward or four reverse selections into play instantly . . . non-stop . . . without gear clash.

Second good to you? Then you'll want to get the good news on Power Shifting across the complete "band" of speed selections, firsthand, in the field. Your John Deere dealer has a "3020" or "4020" he'd like to leave with you for a day. After the tractor has had its say, get the good news on John Deere Tractor financing terms from your dealer.
*Diesel model

REVERSE 8 7 6 5 4 3 2 1 FORWARD N N PARK

JOHN DEERE
Moline, Iowa

8010, 8020, AND TROUBLE

The first John Deere tractor with more than two-cylinders, since the abortive Dain of 1918, was the huge Model 8010 introduced in 1959. It more or less let the cat out of the bag: The reign of the Deere two-cylinder might be nearing its end. Deere had been keeping that fact a strict secret while multiple-cylinder engines for the "New Generation" tractors were being developed. Deere marketing people were rightly worried that if word got out prematurely, sales of the last two-cylinder tractors would suffer.

The John Deere 8010 caused much amazement when it was introduced at Deere & Company's field day in Marshalltown, Iowa, in 1959, because of its huge size: It was 20 feet long, 8 feet wide, and 8 feet tall. It weighed 20,000 pounds without ballast and 24,000 pounds with liquid in the nearly 6-feet-high tires. The 8010 ran with a six-cylinder,

two-cycle GM 6–71 "Jimmy" diesel engine of 215 horsepower (no other Deere tractor had yet exceeded 80 hp), and it had a nine-speed transmission when the most any other Deere had was five. Instead of mechanical brakes, like every other Deere, the 8010 had air units.

Unfortunately for Deere marketing people, the 8010 didn't sell as expected. First, it had unresolved engine and transmission problems. Only 100 production models were built between 1960 and 1961. Just a single 8010 was sold in 1960, and the rest weren't off the books until 1965. Nearly all of the 100 production models were recalled, corrected, rebuilt into 8020s, and then returned to their owners.

The question is: Why were so few units of such an advanced tractor sold? After all, the 8010's unveiling at Marshalltown generated a positive response. True enough, but one big

problem remained: the price. In the early 1960s, $30,000 for an 8020 with a 3-point hitch was a lot of money, particularly when one considered that the 4020, Deere's next-biggest tractor, cost a mere $10,000.

Some of those who did farm with the 8010/8020 liked them and put them to good use. Others said the tractor was "a dog" and complained that its power didn't live up to claims. Probably due in large part to the peculiarities of the two-cycle GM engine, farmers accustomed to the pleasant exhaust note of the new Deere two-, four-, and six-cylinder engines probably didn't run the "Jimmy" hard enough. The engine required mostly wide-open operation, and because it was a two-cycle, it sounded like it was turning twice as fast as it really was. It was a good and efficient motor, but ended up with the unfortunate nickname "Howling Jimmy."

Deere's 8020 articulated tractor was based on the 8010. In fact, many 8020s were rebuilt 8010s that had been recalled by Deere because of serious transmission problems. All 8010s that were recalled and repaired were returned to owners and dealers with "8020" badging. The 8010 design dated to 1959, with the follow-on 8020 showing up in 1963. 8020s had an upfront provision for the mounting of a 'dozer blade, with front-end reinforcement that allowed for the mounting of a blade's lift cylinders. Of the 100 8010/8020 units built, about 75 remain registered with collector groups.

THE "NEW GENERATION"

With secrecy that nearly rivaled the hush-hush status of the Manhattan Project, select Deere engineers were pulled from their assignments and sequestered in what had been a supermarket. What followed was one of the best-kept non-governmental corporate undercover operations in history. The Deere "Butcher Shop Boys" were assigned the task of designing a "New Generation" of tractors with all-new multi-cylinder engines. (There were only two exceptions, and they were for export: The Model 730 two-cylinder diesel was built in Waterloo until early 1961, and continued to be built in Argentina throughout the 1960s.)

On August 29, 1960, the New Generation John Deere tractors were unveiled with great hoopla at the Coliseum in Dallas, Texas, at an event dubbed Deere Day in Dallas. Deere flew in 6,000 dealers, plus press people, in more than 100 airplanes from all across the country. Fireworks, barbecues, and big-name entertainers supported displays of 136 new tractors and 324 other pieces of equipment.

The new line of Dreyfuss-styled multi-cylinder tractors was an unqualified success. The dealers liked them and the farmers bought them. Sales for 1961 and 1962 were up dramatically. The new line had four models: The 1010 at 30 drawbar horsepower; the 2010 at 40 drawbar horsepower (the 1010 and 2010 were built at Deere's Dubuque facility); the 3010 at 55 drawbar horsepower; and the 4010 at 80 draw-bar horsepower. Most were available with gasoline, diesel, and LPG (liquid petroleum gas) engines. They also were available in a variety of configurations, from utility to row-crop. The 1010 was even available as a crawler.

The 2510 of 1965–68 was designed for farmers whose needs fell between the capabilities of Deere's 2020 and 3020 models. And indeed, the relationships were very direct: When a 3020 chassis was altered and fitted with a 2020 engine, the 2510 was born. It was sold in gasoline and diesel versions, as well as row-crop and hi-crop iterations. Both engines ran with four cylinders, with the gas version displacing 180.4 cubic inches, the diesel 202.7. Buyer's choice of a Synchro-Range or Power Shift gearbox provided eight forward speeds. Base price for a row-crop gas-powered model was about $3,975, with diesel, Hi-Crop, and the Power Shift options adding to the bottom line.

When the 4010 took to the fields for the first time in 1961, it was the biggest non-articulated tractor John Deere offered. One of the New Generation models, it was powered by an inline six-cylinder engine that could be ordered with gas, LPG, or diesel fuel options. The gas and LPG engines displaced 302 cubic inches, while the diesel was listed at 380. More than 75 horsepower was on tap, another first for Deere. An eight-speed Synchro-Range gearbox could move the nearly five-ton tractor along at 14.3 mph. Henry Dreyfuss and Associates designed the cockpit and seating, in conjunction with an orthopedic specialist. The seat was the most comfortable yet installed on a Deere.

Visitors to the 1961 Minnesota State Fair were drawn to what fair organizers dubbed "Machinery Hill," where the Deere display showcased the 3010, 4010, and other New Generation tractors. Like any good state fair, the Minnesota event offered stock car and open-wheel racing, thrill drivers, parades, good eats, the appliance-oriented "Electric City," and the Fly-O-Plane and other rides. There also was a midway complete with sideshow performers (don't miss Lobster Boy or Rubber Skin Man), and Club Lido, where scantily clad dancers entertained the grownups. It was in this environment of lively fun and hardworking people that Deere prospered.

In 1960, Deere Day in Dallas attracted about 6,000 dealers and others to a sales and marketing program highlighting Deere's "New Generation of Power"—a completely updated tractor line that had been in the works for a long time. When the consumers spoke later, the New Generation became a tremendous success.

As farmers spent more and more time in their tractors, the need for additional safety and convenience became apparent. By adding an enclosed cabin, as on this 1979 Model 3130 (*both photos*), Deere shielded operators from flying debris, dust, and other day-to-day hazards of working a field. The cabin also added structural strength that helped to protect the driver in the event of a rollover. Many modern Deere models equipped with cabins have air conditioning, audio systems, and GPS systems that provide the farmer with comfort and information.

Evolution being what it is, the tractor world produces many new models that closely mimic their predecessors. The 3020 (*shown*) was a kissing cousin to the 3010 when it came from Deere's Waterloo facility in 1964, but it was more desirable because of additional power, traction lock, and the available PowerShift transmission. For even more convenience, the 3020 gained optional Front Wheel Assist in 1968, and could be had with orchard-friendly sheathing.

Manufacturing plants in Europe have given Deere a global reach for decades, and produced models not commonly seen in the United States. One of those was the 2120, with engines and transmissions (eight forward gears and four reverse) manufactured in Mannheim and Baden-Wurttemberg, Germany, for use in tractors sold across Mexico beginning in 1971. Mexican law mandated that 60 percent of the tractors' materials be supplied locally, so the 2120 was a true American-European-Mexican hybrid. It ran with a four-cylinder engine that was given an added boost with turbocharging.

Deere Portfolio
New Generation Tractors

The 1963–66 Model 4020 was one of the most popular New Generation Deere tractors. Its six-cylinder engine (gasoline, diesel, or LPG) was mated to an eight-speed Power Shift transmission; an automatic trans was optional. Tested horsepower at the PTO shaft was 95.8.

A six-cylinder diesel displacing 531 cubic inches powered the 1966–72 5010. At the rear, the 5010 had a sophisticated Category 3 three-point linkage. Optimal engine performance came at 2200 rpm.

The "Old Style" 2640 (*pictured*) was introduced in 1975, as the latest in a "JD" series that stretched back to 1962. The back-sloping hood design was carried over from the smaller 30 and 40 series. A six-cylinder engine produced 80 horsepower that the operator could put to work via six forward gears. Within the series, the 2040 and 2240 were imported from Deere's Mannheim, Germany, plant while the 2440 and 2640 hailed from Dubuque, Iowa. Any of the four could be ordered in Orchard or Vineyard trim.

In 1977 John Deere introduced the Iron Horse line of tractors, which was marked by upticks in power and strength. The 4440 (*pictured*) was one of these new models, running with a turbo diesel that produced 130 horsepower at the PTO. Options included the Quad-Range gearbox with 16 forward gears and six reverse, or a PowerShift version with eight and four. Two- or four-wheel drive were available, and the enclosed cabin was a comfortable place to spend the day.

As good as the 55-horsepower 4010 was, the 4020 trumped it and became one of John Deere's greatest models ever. The 4010's incredible power was increased, to 94 hp, and the cockpit was redesigned for driver comfort. The buyer now was given the choice of the eight-speed Synchro-Range gearbox or the latest Power Shift. The latter system's single lever was seen as a great convenience, although it did take away a few of the 4020's horsepower in the process. 4020s built for 1969 and beyond had even more features, including a new control panel and an improved 12-volt electrical system with an alternator. It's not difficult to see why 4020 models purchased today cost far more than they did when new.

The 1020 came on line in 1965, and was one of Deere's "Worldwide" designs, with various examples coming from Dubuque, Iowa, and Mannheim, Germany. A three-cylinder engine could be fed by diesel or gas according to buyer preference. The gas version displaced 135.5 cubic inches while the diesel came in at 152. The direct-injection diesel eliminated glow plugs that prolonged the starting process. Eight forward gears and four reverse propelled the 1020 to 17.1 miles per hour in high gear, and 5.1 mph in reverse. Weighing a shade over two tons, the 1020 carried its fuel up front for better distribution of poundage when loaded.

GOING INTERNATIONAL

William Hewitt's time as president brought the company onto the international stage. He established operations in Mexico and in Central and South America. Under his leadership, the German company, Heinrich Lanz, was acquired, along with Lanz's famed "Bulldog" line of farm tractors.

In the early 1960s, several of the leading agricultural equipment companies created standardized products to be built by their international operations. In 1963, Deere & Company followed suit. Two years later, three new "Worldwide" tractors were introduced. This trend has been expanded since then to include the entire tractor line.

The 730 was another machine designed by Deere to cover the gap until the New Generation units were ready for sale in the summer of 1960. Sold in row-crop, single front wheel, hi-crop, and standard tread versions, the 730 was versatile but had only six forward speeds and a solitary reverse. Most 730 production was in Waterloo, Iowa, but some of the tractors were sent in kit form to Rosario, Argentina, and Monterrey, Mexico, for final assembly. All of the Argentinean models were electric-start diesels, while most Monterrey versions were row-crop diesels.

Deere's "worldwide tractor" project was unveiled in 1965 with the 20 Series that included the Model 2020 (*pictured*). Because the 20s were intended for use in America and abroad, several variations were available. The four-cylinder 2020 was built in Dubuque until 1967, when assembly shifted to Deere's Mannheim, Germany, and Saltillo, Mexico, facilities. This 2020 is sheathed in an orchard package that encloses the hood and rear wheels to protect trees during operation. Even the operator's pillion has been cloaked in a protective screen that allows forward visibility without danger to driver or trees.

The Rusty Palace

The "Rusty Palace" is the informal name given to the Deere & Company Administrative Center in Moline. Deere president William Hewitt recognized that the architectural excellence of a new headquarters would encourage employees to be "bold, ingenious, and creative," and would also influence how customers would regard the company. Hewitt surprised many by tapping noted Modernist architect Eero Saarinen to design the building. His challenge to Saarinen was bold and inherently contradictory: Design a thoroughly modern complex befitting down-to-earth, rugged men of the soil.

The architect responded with a building consisting of three sections stretched across a richly landscaped valley. These primary units house the main office, the auditorium-product display section, and the west office wing. The external structure, with exposed beams designed to carry a thin coating of rust, employs an agent that limits corrosion. Glass walls provide views of the beautiful Illinois landscape. Excited Deere employees occupied the new building in 1964.

Deere & Company's award-winning world headquarters building, which opened in 1964, stands amid 1,400 acres that also host deer and geese. The complex was designed by Eero Saarinen, the Finnish-American architect who also developed the Gateway Arch in St. Louis. The decision by then-Deere president William Hewitt to build the new headquarters in Moline was a saving grace for the city, which had feared the loss of the company's home to a major city on the West or East coast. Saarinen died before construction began.

Chamberlain Holdings Ltd. dated to 1949, and was Australia's only domestic tractor company. The 1965 Super 90 was powered by a three-cylinder diesel engine that displaced 212.7 inches and operated at 1200 to 500 rpm. Nine forward gears and three reverse gave the 6-ton tractor a useful versatility. With Deere's acquisition of 49 percent of the company in 1970, Chamberlain-John Deere Pty. Ltd. was formed. Much like the earlier Lanz purchase, this partnership gave Deere another venue in which to sell tractors without having to start from scratch.

It was March 1972 when Deere replaced the 5020 with the Generation II 6030 (*pictured*). The 5020 was one of several models to be upgraded at this time, and the 6030 was the first Deere tractor to offer the buyer a choice of engines: a 404-cubic-inch, 146-horsepower unit or a 531-cubic-inch, 175-horse motor. Both were six-cylinder turbos with intercoolers. An enclosed Roll-Gard cab could be ordered, along with dual wheels at the rear for added grip. The 6030 was sold through 1977.

Deere president William Hewitt (*left*) and architect Eero Saarinen examine a tabletop model of the proposed new Deere headquarters.

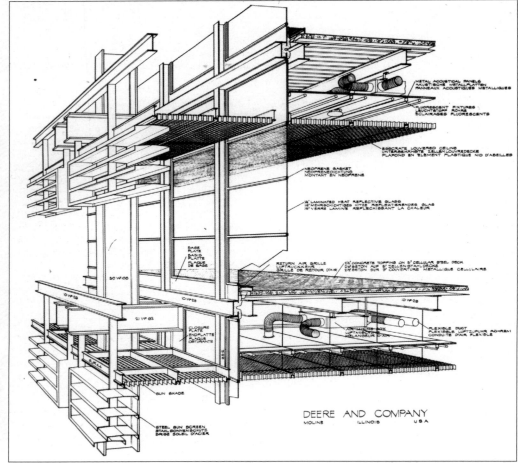

This cutaway rendering emphasizes Saarinen's dramatic use of steel beams.

The building complex dominates a landscaped valley. Hewitt wanted a design that was modern, but "down to earth and rugged" as well.

The strikingly geometric scheme is apparent in the interior (*left*). The whole complex (*above*) is at once Modernist and plainspoken.

INDUSTRIAL EQUIPMENT AND CONSUMER PRODUCTS

Deere's Industrial Equipment Division was established in 1956, along with a separate dealer and marketing organization. After that, a separate engineering department was added. In 1962, a completely new line of products was unveiled: the JDs. As you might guess, all model designations began with the letters "JD." The line now included motor graders, logging devices, loaders, backhoes, excavators, and crawlers. The division now challenged Deere's old friend, Caterpillar, in all quarters.

Caterpillar wasn't Deere's only rival in this segment. To the contrary, Deere simply wanted a piece of a growing pie. International Harvester and Case had complete lines of construction equipment, and there was foreign competition from Komatsu, Kubota, and Mitsubishi.

In 1963 a new Deere Consumer Product Division was established at the old Van Brunt factory in Horicon, Wisconsin. Its first product was the immensely popular Model 110 lawn tractor. The line developed to include, at least for a time, bicycles, snowmobiles, all-terrain vehicles, and chain saws, but now concentrates on small tractors, mowers, snow blowers, and other items for the homeowner.

Extensive effort is made to ensure that every Deere product is safe. In this unusual factory image snapped in about 1963, chains suspend a Model 110 lawn and garden tractor to test for rollover tolerances. By learning the extremes needed to cause a tractor to tip, Deere engineers eliminate potential problems before the public ever sees the product.

By the 1950s Deere tractors had become common in the logging industry. The 440-C (*both photos*) was adapted from the 435 for that use, and was seen with wheels or crawler treads. Because of the rugged nature of logging, many of the 440's parts were upgraded for durability. A cast iron radiator grille and extra-sturdy hood helped to protect the engine from damage. For additional security, filler caps were moved under the hood. Instead of the appliqués worn by most John Deere tractors, the "John Deere" name was formed into a steel plate affixed to the sides of the 440-C. A 106-cubic-inch, two-cylinder diesel resided under the reinforced hood and was coupled to a four-speed gearbox. PTO horsepower was 32.91, and enabled the 440-C to confidently move earth and large sections of fallen timber.

In the mid-1960s Deere touted its mammoth Model 105, shown here at the Minnesota State Fair, as the "world's largest" combine harvester. The model was emblematic of the company's success: In 1963 Deere became the world's No. 1 producer and seller of farm and industrial tractors and equipment, passing International Harvester. Just as important to farmers, however, were innovations smaller than the 105. The development of a two-row corn head in 1954 for Deere's Model 45 combine allowed a farmer to harvest as many as 20 acres of corn a day with one machine.

By the mid-1960s Deere was well established in the home-products segment. This 1967 advertisement suggests that a Deere lawn and garden tractor would be the ideal Christmas gift—and if you had a substantial lawn, you'd probably agree. Note how the kiddie products at the ad's lower left broaden the Deere mystique to include every member of the family.

The JD646-B compactor (bottom) and JD860-A scraper (top) of 1975 were two more Deere machines designed for non-farm use. A compactor's primary task is to move large amounts of material, such as landfill waste, from one location to another. Deere's JD646-B compactor had wide steel wheels with angular teeth that gave terrific traction without getting mired in mud or other debris. The front bucket could push very large quantities in a single load. The paddlewheel mechanism of the scraper processed and distributed material in conjunction with the compactor, or could work solo in grading operations. As suggested by this ad, this heavy-duty pair was pitched as a team.

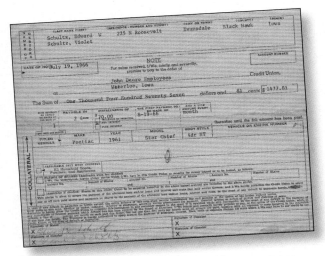

At a glance, Deere's attitude toward its employees might be perceived as paternalistic, but that would be a misinterpretation. Rather, the company treats its people with respect, and offers practical aid, as well. This 1966 document relates to Deere & Company financial assistance for an employee who was buying a used car.

"Charge it"—a clever double meaning suggesting that payment plans were available to purchasers of Deere batteries, chargers, tools, and other everyday home items.

JOHN DEERE
MEMORABILIA

Drawn and Integral Disk Harrows

Whether pulled by the tractor or an integral part of the machine itself, disk harrows of the 1960s and '70s were designed to handle a full range of soil types and crop requirements.

JOHN DEERE DISPATCH:

A "Power Train '66" advertisement highlights Deere's all-new Model 2510, which had bowed in the summer of 1965. A row-crop variant is seen here.

NEW DELUXE ROW-CROP TRACTOR WITH 53 H.P.

HYDRAULIC CUSHIONING

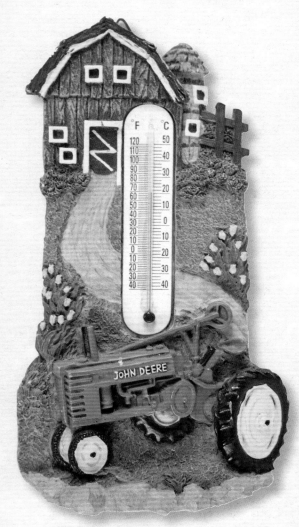

A wall-mount Deere thermometer isn't as whimsical an item as it may seem, for what's more important to a farmer than the weather?

CORN PICKERS and SNAPPERS

Corn snappers, which simply harvest the ears of corn, were still popular in the 1960s but are seldom seen today. Corn pickers, which simultaneously harvest and shuck the ears, are more useful.

The enormous expense of maintaining a farm operation is apparent in this ad from about 1970: Combines were huge machines that easily cost six figures. A combine-harvester like this one was set up for both cutting and threshing.

STANDING GRAIN AND SORGHUM COMBINES

Part promotional gimmick, part currency, the Deere Dollars of 1979–81 were printed by the American Bank Note Company, and were given away (and taken back from customers, for credit) by Deere dealers.

"Action" toys became the rage in the 1950s, and by 1974 every kid wanted something that could be handled, pushed, pulled, or otherwise put to work.

In 1972 Deere & Company issued a cookbook with recipes contributed by Deere employees. The collection was so popular that Deere reissued it in 1976.

The End of Family Management

In 1975, William Hewitt, with an eye to the future, created the position of executive vice president and named Robert A. Hanson to fill it. Hanson had come up through the ranks, spending many years in foreign operations. In 1982, when Hewitt reached the mandatory retirement age of 67 (a policy  he himself had instituted), Hanson was promoted to chairman and chief executive officer. With Hewitt's retirement and Hanson's appointment, 145 years of Deere family leadership came to an end.

Robert A. Hanson retired in 1990 to be replaced by Hans Becherer, another long-time Deere man who had worked his way up through overseas divisions.

A new regime, new and better products.

A 1979 visit to John Deere headquarters in Moline by a trade delegation from the People's Republic of China was part of a long history of Deere efforts to do business in that country. As early as 1912, the company's San Francisco branch manager toured China. Immediately after President Nixon's overtures to China in 1972, Deere leadership and Chinese officials began a series of reciprocal visits that led to the importation of Deere tractors and technology to China in the 1980s. Since then, Deere has entered into significant joint ventures with the Asian giant.

DEERE R&D, AND A FLING WITH ROTARY

Deere has routinely allocated a greater portion of its profits into research and development than most such companies. Technically advanced new products are thoroughly proven before being released to customers. The Deere Research Center, for example, has facilities for testing implements and tires in various kinds of soil. There's also a Cab Simulator that rivals an aircraft flight simulator. It can be programmed to give the operator the exact sensations encountered in operating, for example, a combine on a particular sort of terrain.

The simulator was one of many innovations and experiments developed by the research center. Few people outside of Deere know that the company once owned the Western Hemisphere rights to the Wankel rotary engine. (Japan's Mazda held the

At the University of Nebraska at Lincoln in 1982, Test #1458, of a Deere Model 4250 tractor (15-speed Powershift diesel), was carried out at the UNL test track. This and all other UNL tests followed the Agricultural Tractor Test Code approved by the American Society of Agricultural Engineers and the Society of Automotive Engineers. The silver vehicle seen here is a tractor test car that was hooked to, and reported on, the 4250's vital functions. PTO and drawbar power were evaluated; likewise fuel consumption and the tractor's performance during two hours at maximum engine power.

rights for the Eastern Hemisphere.) These rights were purchased from the Curtiss-Wright Corporation, which had been developing the engine for military tank use. For 15 years, beginning with the fuel-crisis year of 1973, Deere worked on the engine for possible use in tractors and combines. Deere's interest stemmed from concerns similar to those that prompted the transition from the two-cylinder engine: weight and fuel consumption. In this case, it wasn't the amount of fuel burned, but the type of fuel. The Deere rotary incorporated "stratified charge" fuel induction, allowing it to burn just about any combustible liquid without adjustment. The idea was that, in the event of another fuel crisis, farmers could burn diesel, gasoline, furnace oil, or even homemade alcohol in their rotary-engined equipment.

Interestingly, at least one company brochure from the 1980s was clearly aimed at military applications (Curtiss-Wright all over again!). One page of that brochure is dominated by an illustration of a tracked, armored vehicle in military green. Above it, a bold headline declares, "PLAN NOW TO INCLUDE SCORE™ ROTARY POWER."

This was provocative stuff, but by about 1988 Deere felt it had reached the end of the road with rotary-engine research. The company sold the rotary program's physical assets and intellectual property to Rotary Power International, of Woodridge, New Jersey, early in 1992. RPI later guaranteed a minimum of 10,000 hours of engine operation without overhaul, but for Deere, the long dalliance with rotary power was over.

Deere has pursued and won military contracts, but made a unique push for them in the 1980s, when the company explored rotary engines. Rotaries are simultaneously simple and complex, and although Deere committed many resources to research, the rotary program was sold off in 1992.

Above: Deere had offered operator cabs on its tractors since the late 1950s, but in 1976 the company was granted a patent for a new "vehicle roll-over protective structure"—in other words, a safety cab that protected the operator's life and limb in case of the most dreaded of tractor accidents. A subsequent advertisement (*above, right*) suggested that although a Deere cab could be damaged, the driver would walk away.

WORLDWIDE CHALLENGES OF THE 1970s AND '80s

In the early 1970s, the Soviet Union made a massive grain buy from the USA that disrupted the normal equilibrium of farm supply and demand. Then came the Arab oil embargo of 1973, which tripled the price of American fuel. This was followed in 1980 by a U.S. grain embargo imposed on the Soviets in retaliation for their invasion of Afghanistan. This action was another tremor that rocked the usual forces of agricultural supply and demand. On top of these developments was Washington's general mismanagement of the U.S. economy, which led to interest rates as high as 18 percent. American farmers, and farm equipment companies, were in real trouble because both relied on the free flow of reasonable credit.

When inflationary pressures were brought under control, land values and farm prices fell. By 1984, the plight of the American farmer was desperate. Foreclosures and auctions were common, and the entire equipment industry suffered. International Harvester was forced into a merger with Case. Ford purchased New Holland and Canadian Versatile, but ultimately sold its farm-equipment operations to Fiat. Massey Ferguson, White, Allis-Chalmers, and others joined forces and regrouped under the AGCO banner.

In the late 1990s, the already-merged I-H and Case became part of CNH (Case-New Holland) Global, with Fiat holding a majority ownership stake. Only the Illinois neighbors, Deere and Caterpillar, have managed to survive as independent operations.

Electrician and farmer Buddy Vidovich, a member of the Native American Paiute tribe, pulls a John Deere seeder attachment at a site near the Truckee River in western Nevada. The Paiute are descendents of some of the earliest known inhabitants of the Great Basin desert region in the West.

In the late 1970s, Deere introduced a pair of related, self-propelled forage harvesters, the "compact" 5440 (*background*) and the larger 5460. These were heavy-duty machines designed to cut and chop numberless rows of stalks and other forage into silage suitable as feed for livestock. An oil-cooling system and versatile clutches helped protect the gearcases from burnout. Because the tungsten-carbide cutterheads on both of these machines were fully two feet in diameter, many acres could be cleared in a day. "Personal-Posture" seats, and optional air conditioning and four-wheel drive, made the work a little easier.

MORE POWER FOR YOU—AND THE GUTS TO BACK IT UP
NEW 178-hp AND 255-hp JOHN DEERE SP FORAGE HARVESTERS OFFER HEAVY-DUTY, PROTECTED DRIVES TO MATCH POWER OUTPUT

THE FORAGE SPECIALIST

Even something as sturdy as a Deere machine eventually reaches the end of its life. If it can't be rebuilt, it likely ends up in a place like this: a farm-machinery graveyard. The assortment of tractors, combines, and implements may strike a melancholy chord, but parts salvaged from discarded machines contribute to a thriving secondary-market industry, which sells them as-is or as rebuilds.

THE GRAIN EMBARGO

In 1979 the Soviet Union invaded Afghanistan in order to prop up an unpopular Communist government that had overthrown the Afghan monarchy. In Washington, President Jimmy Carter responded with a suspension of all grain deliveries to the Soviet Union other than the eight million tons guaranteed under a bilateral agreement made in 1975. This included feed grain and wheat, as well as soybeans and meat.

Carter's assumption was that the USA had "food power" over the Soviets, and that this punishment would force a speedy departure from Afghanistan. Carter was dead wrong: As far as Moscow was concerned, politics and strategic influence trumped grain. The U.S. embargo remained in place until the first months of Ronald Reagan's first term—16 months in all—and the defeated Soviets didn't stagger from Afghanistan until 1989.

In the meantime, the people most hurt by the embargo were America's farmers, for two reasons: First, they had a long history of annual grain surpluses, and the Soviet Union had been a major purchaser of it. When the embargo prevented much of that surplus from being sold, the price of grain fell.

Second, many farmers who had borrowed heavily to expand their businesses earlier in the decade were now unable to meet their debt. Although Washington spent $2.2 billion to purchase canceled grain contracts, growers continued to suffer.

In a sort of agricultural domino theory, strapped farmers had to postpone or completely forgo purchases of new tractors, harvesters, and other products manufactured by Deere and others.

For Deere, the negative effects echoed for years. In 1986 the company posted a loss of $229 million, its first since 1933. In 1987 Deere ended the year in the red yet again, this time to the tune of $190 million. Happily, as the 1980s wore into the '90s, Deere's continuing investment in R&D paid off with handsome profits.

A charming, good-sized wood representation of Deere's pre-tractor days, when strong teams pulled Deere wagons around America's farms.

Deere introduced what would be a 12-year snowmobile program in 1972. The cartoon ad from 1976 was done by animation artists at Hanna-Barbera, which produced at least one snowmobile safety film for Deere. By 1984, the program's last year, the machines ran with Kawasaki engines. All production took place at the John Deere Horicon Works in Horicon, Wisconsin.

HOW TO LET PEOPLE KNOW YOU'VE ARRIVED BEFORE YOU EVEN GET THERE.

Admittedly, we're very proud of the new John Deere Sportfire's exterior.

But the exterior improvements came after the interior improvements.

Before we attached the speedometer outside, we attacked the powerplant inside. And installed a new 436cc engine. With a more efficient expansion chamber to boost both the horsepower and the miles per hour.

Before we chromed the soft-ride shocks, we took a hard look at the entire suspension system. And gave it a John Deere aluminum slide-rail system that actually makes moguls seem flatter.

And before we designed the new rally striping to make the Sportfire suit your eye, we designed an action-formed sport seat to make it suit your seat.

The new John Deere Sportfire™ Designed inside and out to look as good standing still as it does at top speed.

For more information, write John Deere, Dept. A53, Moline, IL 61265.

RIDE THE NEW PRIDE OF DEERE

In the manner of toys from decades past, this Deere truck from about 1980 is made of tin, with the added usefulness of a trailer that could be opened and filled with candy, stray buttons—you name it.

Among the more peculiar Deere products is this interior-lighted tractor from about 1984. Approximately 10 inches long, it's molded from a single piece of nubbly, highly flexible plastic, and was apparently intended as a display piece.

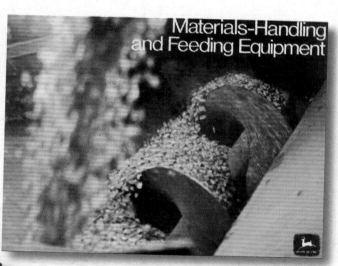

This ad from the 1980s, with its striking detail photography, emphasizes how Deere machinery handled a farm's feed.

Materials-Handling and Feeding Equipment

Some of the nicest pieces of Deere memorabilia are homemade for personal use, such as this friendly "Welcome" sign made from wood.

This numbered, plastic and tin truck from the 1980s doubled as a bank, while recalling the 1920s.

The Midwestern winter of 1978–79 was unusually ferocious, so this '79 ad, with its rhetorical "Remember last winter?" headline, undoubtedly helped sell a lot of Deere 320 snow throwers.

Log harvesting became an easier job last summer when we introduced our new JD640 Grapple Skidder. That's not just because the grapple eliminates most of the difficult work of attaching cables to trees and sawlogs. We also put operation of the grapple under remote pushbutton control. The operator stays in his protective compartment while he picks up the load. Rounding out our line to six skidders, this one also marks our tenth year of engineering and marketing these modern 4-wheel-drive logging machines.

Your inquiries about John Deere products for forestry and construction and about the company that builds them are welcome.
John Deere, Moline, Illinois 61265.

JOHN DEERE on the move

23

Remember last winter?

Be ready this year with a new compact John Deere Snow Thrower

Once upon a time, no kitchen or wet bar was complete without a wall-mounted bottle opener. The Brown Manufacturing Company of Atlanta made this Starr-brand example with Deere markings.

Forestry developed into a big product area for Deere, as suggested by this aggressive 1975 ad.

TRACTORS LARGE AND SMALL

Despite hard times worldwide, and the highly visible disappointments of the 8010 and 8020 models, Deere regrouped and forged ahead with development of its New Generation of articulated tractors. In the interim, Deere contracted with Wagner Tractor, Inc. of Portland, Oregon, to manufacture a pair of articulated models, the WA-14 and WA-17, that acted as placeholders until the Deere nameplate showed up on the articulated Model 7020 in 1970. This improved and enlarged New Generation of articulated design, which grew to include the 7520, ran with turbocharged and intercooled engines, and flourished until the late 1980s, when yet another articulated-design generation was introduced.

Deere remained mindful of the need for smaller tractors, as well, recognizing in the early '70s that there was a growing market for machines in the 20–30-horsepower class. Farmers wanted these specialty machines but Deere had been slow to service that need. Further, a market survey indicated that if Deere were to produce tractors in this class in the USA, the tractors couldn't compete on price with Japanese brands. Therefore, a deal was struck with a Japanese firm, Yanmar, to build tractors for this market, to Deere specifications. Deere soon had a competitive line of diesel-powered tractors with power ratings that ranged from 11 to 30 horsepower.

Innovation continued into the 1980s. A new loader/backhoe that was introduced in 1983 spurred sales in Deere's Construction-Industrial Division. In agriculture, Deere's new MaxEmerge planter allowed farmers to place seeds in appropriately spaced rows at the proper depths in challenging field conditions. Deere's Model 900 combine helped the company garner a 40-percent worldwide market share in that segment. Another advance was the Hydrapush, a new concept in manure spreaders that utilized the tractor's hydraulic power to force the manure back to the spreader's flippers. The Iron Horse tractor series of 1977 was dramatically improved for the '80s. Refinement also came to Deere's 50, 55, and 60 Series tractors. Finally, as befits its history, Deere continued to develop a variety of innovative tractor-mounted plows.

At the end of 1986, Deere & Company was poised to enter its 150th year of existence, making it one of the oldest firms in the world. As Deere stood at this juncture, it was unique in the John Deere "attitude": From the company's mahogany offices to the production floor, there was a universal, almost nostalgic respect for the past, and a sense of excitement about the future.

As farm operations grew larger to meet market demand, farmers' desire for additional power became a roar. Until the mid-1960s 75 horsepower was deemed adequate for most uses, but even that lofty output was about to go by the wayside. Part of the New Generation line, the 5010 set new standards for Deere power when it was introduced in 1966. With a six-cylinder diesel under the cowl and an eight-speed Synchro-Range gearbox, the 5010 churned out 121.12 horsepower at the belt. As Deere's first two-wheel-drive model to exceed 100 horsepower, it was welcomed by farmers with large properties. The 5010 was available in standard tread and diesel only. When new, the tractor carried an MSRP (manufacturer's suggested retail price) of $10,730 and weighed more than 17,000 pounds when loaded for work.

In 1982, following on the heels of the diminutive Model 650 (at 1,530 pounds, the smallest farm tractor Deere had ever built), the company rolled out its largest to date: the 37,480-pound Model 8850. It was the biggest of a four-wheel-drive series that included the 8450 and 8650; all three were articulated (center-hinged) to allow for tight cornering. The 8850 ran with an eight-cylinder diesel that displaced 955 cubic inches and produced 304 PTO horsepower. Given the machine's immense weight, it's no surprise that it drank 19 gallons of fuel during every hour of PTO operation. Sixteen forward gears and the eight-wheel option seen here made the 8850 a real workhorse.

Yanmar of Japan was founded in 1912 and made its name with diesel engines. It entered the agricultural-equipment business in 1961, bringing its diesel technology to a successful line of small, modestly powered tractors that established a presence in the United States. For Deere to design and build from-scratch models that were similar to Yanmar's would have been eco-nomically infeasible, so in 1977 a co-op deal was struck by which Yanmar would restrict the sales of its branded tractors to overseas markets, while Deere would sell them stateside and in Canada as Deere machines. The 20- to 40-horsepower diesel models came to North America adorned in the familiar green and yellow livery, and sold well.

When Deere wanted to establish a presence in the small-tractor segment in the 1970s, it contracted with Yanmar of Japan to build the machines, using three-cylinder Deere engines. This 1978 brochure highlights the 27-PTO-horsepower 950.

1987–2000
The Modern Company

"Deere has gone around, and actually to the detriment of their balance sheet, given support to their dealers. They are the only [farm machinery company] who will end up with a healthy dealer network."

—*The Wall Street Journal*

150 YEARS IN PERSPECTIVE

From one man, John Deere, who developed the world's first commercially successful self-scouring steel plow in 1837, came the multi-billion-dollar international Deere agri-business of 1987. In the annals of American business, it was an estimable achievement.

Deere has been unusually well managed for many decades, but one doubts that John Deere, or any of the company leaders after him, until perhaps William Hewitt, could have had any inkling of what the company would one day become.

The growth of Deere parallels the development and settlement of the Midwestern United States, an area that early pioneers called the "Golden Land of Promise." The time period encompassing 1837 to 1987—the life of one man, John Deere, to the dominance of

In July 2000 Deere & Company updated the familiar Deere logo, unveiling the eighth version in the company's history. Chairman Han Becherer (*at left*) and President Robert W. Lane announced the update at company headquarters in Moline, Illinois. From the earliest trademarked version of the leaping dear (1876), Deere marketers took pains to match the logo to the company's strategy. As Deere branched out in the 1950s and '60s into products for the home market, for example, the words "Quality Farm Equipment" disappeared from the logo. The latest version (*pictured*) gives added prominence to the company name and, for the first time, shows the deer bounding upward instead of down.

Mindful of the ever-increasing demands of modern farms, Deere introduced the 8100 in 1994. Built on a 116.1-inch wheelbase and powered by a six-cylinder turbocharged diesel, the 8100 was nothing to trifle with. The 466-cubic-inch motor delivered 160 PTO horsepower and offered the operator 16 forward speeds and four in reverse. Inside the enclosed cabin you might find air conditioning and an AM/FM/cassette stereo system to provide some comfort while working the soil. Sold in a choice of two- or four-wheel drive, the 8100 weighed nearly 18,000 pounds when prepped for use. The tractor could carry 135 gallons of fuel, enough to provide a full day's work from a single fill-up. The 8100 was built through 1998.

John Deere, the company—is highlighted by people, progress, and products that reflect that company's commitment to integrity, quality, and innovation.

MANAGEMENT STYLE PAYS OFF

Modern-day CEOs of Deere & Company, while no longer part of the John Deere family tree, are unique among leaders of major American businesses. *Management Today* magazine noted that Deere managers don't fit the profile of career-oriented managers who consider one company as useful to his or her career as any other. To the contrary, "the Deere men believe in their company, and not just out of expediency. They honestly feel that Deere deserves their fealty and reverence."

Robert A. Hanson's tenure as Deere CEO began in 1982. The general economy still wasn't completely healthy, and the Deere dividend to stockholders was cut. But by 1989, Hanson autho-

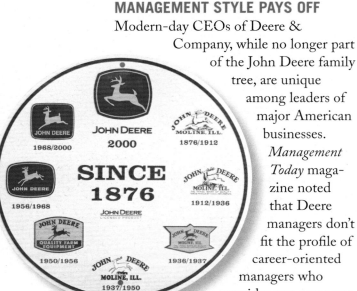

The leaping John Deere buck is one of the most appealing and well-known symbols in all of American business. This commemorative sign displays every example since the logo was registered in 1876. Progressive simplification allowed the logos to be more easily stenciled onto Deere products.

rized the restoration of dividends to previous levels. He also oversaw the largest product introduction in Deere's history: round balers, all-new combines, updated tractors of over 100 horsepower, variable-row-width cotton pickers, and a new series of disk harrows.

The U.S. economy revved up in 1988, and Deere's sales increased by a whopping 30 percent over the year before. In fact, 1988 ended two years of losses with a profit of $315 million, a company record.

Kansas-based Funk Manufacturing Company, a maker of powertrain components for industrial, agricultural, and off-highway use, was acquired in 1989, and gave Deere improved access to and oversight of these critical components.

Hans W. Becherer, who was elected chairman and chief executive officer in 1990 upon Hanson's retirement, had extensive experience in developing Deere's international presence. In 1991, for example, Becherer oversaw Deere's purchase of SABO, a German maker of lawn mowers. Just two years later, Deere's home-and-garden sales surpassed $1 billion for the first time. Becherer would continue to pursue the company's international efforts after taking the big chair.

Large farming operations may employ several John Deere combines to bring in the harvest. The middle combine in this picture has extended its unloading auger and is passing its contents to the auger wagon. The wagons allow the combines to continue harvesting as cargo is taken back to the driers. Many of Deere's bigger combines can hold 300–350 bushels internally before they need to offload.

no new lawn mower ever had a bigger name to live up to

But what's in a name?

Simply this. With John Deere walk-behind mowers, you get the same kind of reliability our big tractors bring to farms and industries. You'll also get a grass-catcher (standard on all six models*). And a dipstick oil level check. And a gas gauge. And quick-release wheel locks that let you change cutting height in seconds instead of fooling around with nuts and bolts. Plus extra-easy starting and approved safety features. Buy the mower that lives up to the John Deere name. We wouldn't offer you anything less.

*21-in. self-propelled with key-electric start/21-in. self-propelled with recoil start/21-in. hand-propelled with key-electric start/21-in. hand-propelled recoil start/19-in. hand-propelled with key-electric start/19-in. hand-propelled with recoil start.

The housing boom of the 1980s and '90s encouraged Deere to get into the home & garden market, with products that included this push mower.

From the Waterloo Boy to the Model D, these palm-sized cast toys encapsulate some 40 years of Deere history.

For many, Deere represents nostalgia for more innocent times. Latter-day Deere lunchboxes with that theme are particularly popular collectibles.

Not buying a John Deere?
Demand ANOTHER $9,000 off

You may need it...
just to cover your losses on resale

You've been offered a deep price cut. You must have been. Otherwise you wouldn't consider less than a John Deere. But before you sign on the dotted line, do two things:

First, demand another $9,000 off the price of the "super bargain" tractor. You may need it to cover your losses when you sell a "super bargain" tractor.

Resale is the hidden loss you probably haven't considered. According to the Fall 1984 Official Guide of the National Farm and Power Equipment Dealers Association, the average 5-year-old John Deere 4440 sells for 108 percent of its 1979 list price. One leading competitor averages 88 percent of its 1979 list price. The other averages 74 percent (see chart). Add the John Deere gain on original price to the competitor's loss. The John Deere advantage is as much as $9,078!

Second, see your John Deere dealer.

Don't assume his price is much higher. Special factory incentives and interest waiver programs are now available. The up-front price difference may now favor John Deere.

Now reconsider. Are you still willing to settle for less than a John Deere? Remember you give up John Deere productivity, fuel efficiency, parts availability, even operator convenience ... as well as resale value. That's a lot to give up. But if you do, make sure they make it worth your while.

See your John Deere dealer for details on your power size: 100-hp 4050, 120-hp 4250, 140-hp 4450, 165-hp 4650 or 190-hp 4850.

Note: Average resale prices taken from Fall 1984 Official Guide of the National Farm and Power Equipment Dealers Association. Percentages calculated from best currently available information on 1979 list prices.

Resale value of equipment can be critical to farms that regularly upgrade. This ad, from about 1988, declares that Deere is everybody's best bet.

The ever-popular Model D adorns this decorative tin tray.

QUALITY FARM EQUIPMENT SINCE 1837

Green Magazine was a longtime labor of love created by Deere enthusiasts. Typical issues, like this one from 1999, looked back at favorite tractors, with original ads, specs, and other information.

Performance

All-new 16-speed Power Shift transmission: Enhanced speed ratios, reliability, and unmatched power-flow efficiency

Here's one of the most efficient power-shift transmissions on the market. In the working range, we minimized gear meshes to ensure maximum power flow. A 21-gpm gear-driven tandem pump supplies the transmission with ample oil to ensure positive lubrication regardless of terrain. As each gear engages, you get immediate cooling and lubrication of the rugged disk and separator plates. Oil continues flowing for a few seconds after engagement to ensure added protection and reduce clutch drag. (An internal pump keeps pump drive, PTO, and rear axle well lubed.)

The gears feature a long-tooth design with high-contact ratios for added strength and less noise. The massive traction clutch is the most robust in the industry. And the thick cast housing, resulting from years of analytical testing, provides a rigid, structurally sound frame member.

Thumb easily through 16 forward, or 4 reverse speeds, without clutching. Computer modulation and electronic engine-load monitoring PLUS a majority of single-element shifts (only one clutch engages to change gears)... help ensure smooth, comfortable transitions.

Forward speeds range from 1.4 to almost 24 mph. Gears in the field working range progress in ⅓-mph increments to let you match conditions perfectly. We also incorporated a unique speed grouping. The lowest four speeds, and four top speeds, increase 27 percent as you shift up. The eight field speeds increase only 13 percent. This provides a wider range of speeds, with no useless gears. This transmission will remember the last gear that you were in, up to 11th. It also matches ground speed if you engage and release the clutch during transport.

Deere's 16-speed Power Shift transmission was an important selling point in the late 1980s. Besides the forward ratios, the trans offered four speeds in reverse—all of them controlled by a simple thumb lever.

HOME, GARDEN, AND ACQUISITIONS

In the meantime, agri-business continued at home, of course, and Deere responded with expansion and reorganization of company divisions producing equipment for lawn and ground care. Since 1970, these combined U.S.-Canada operations had been part of the farm-equipment group, but in 1991 lawn care was spun off as a separate division. That same year, Deere acquired SABO, a German manufacturer (established 1932) of mowers and other maintenance equipment. By 1993, sales of Deere lawn and garden equipment broke through the $1 billion barrier for the first time.

Americans and American industry became increasingly safety conscious in the 1990s. For manufacturers, an emphasis on safety was good for public relations and good for, well, safety. In 1992 Deere inaugurated a program under which older Deere tractors could be retrofitted with rollover-protection cages and seat belts. The program was a culmination of a decision made by Deere back in 1966,

Not every farm task requires a large machine. Sometimes, the farmer simply needs to get around a large piece of property quickly. This is the role of the Deere Gator (*right and far right*), a simple runabout that was available in a wide variety of configurations, including gas and diesel. The bed located behind the seats could carry all manner of supplies and tools around the farm, and even across the grounds of an oversized residential property. Some Gators, painted olive, went to the U.S. military.

In 1963 John Deere began producing a line of home and garden tractors at a factory in Horicon, Wisconsin. Within three decades more than 2 million products had rolled off the Horicon assembly line. A Deere film crew was on hand to document a 1992 parade through the city to celebrate the milestone. In 1993 Deere sales of lawn and garden products topped $1 billion.

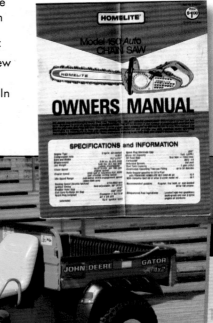

Deere entered the mid- and economy-priced lawn & garden market in 1995 with "Sabre by John Deere," a line comprised of riding and walk-behind mowers. Contrary to periodic rumor, Sabre was manufactured by Deere to Deere specs, and was *not* foreign product with a Deere nameplate. Another brand, Scotts, appeared to be a competitor, but Scotts, too, was owned and built by Deere. Sabre by John Deere was discontinued in 2002, so that Deere could pursue a new arrangement with Home Depot.

Deere continued its history of acquiring successful tool manufacturers in 1994 when it bought Homelite, then a leading producer of handheld outdoor power equipment. Although the move to acquire a line of leaf blowers, generators, and pressure washers fit a Deere strategy to sell to the residential market, the company later reversed course and sold most of the Homelite operation in 2001.

when the company designed the first commercial roll-cage technology for tractors, and subsequently released that design to the industry at no charge.

Smart acquisitions continued. Homelite, a New York maker of chain saws, leaf blowers, trimmers, and other handheld outdoor power equipment, was picked up by Deere in 1994.

A year later, "Sabre by John Deere" brought consumers four mid-priced lawn tractors and a pair of walk-behind home mowers. In an interesting departure from tradition, the Sabre line was sold not just at Deere dealerships, but at national retailers and home centers. By 2010, walk-behind lawn mowers came with what Deere dubbed a "MowMentum Drive System."

The Straight Story

Long after the popular Deere Day short features had faded from the scene, a Deere product played a central role in an Oscar-nominated Hollywood feature in 1999. Director David Lynch's *The Straight Story* starred Richard Farnsworth (*below*), who was very ill during filming, as 73-year-old World War II veteran Alvin Straight, (*left*) who rode his lawn mower 240 miles from Iowa to Wisconsin to visit his dying brother. The real-life trip aboard Straight's 1966 Deere 110 riding mower took six weeks.

Whether in deep winter or high summer, homeowners enjoy the benefits and versatility of John Deere's small tractors. The machines are available in a wide range of models and power levels, and with a great variety of add-on implements that make short work of snow removal or lawn maintenance. This chilly scene unfolded in Boise, Idaho.

STAYING HEALTHY

Deere Family Healthplan centers opened in Des Moines and Waterloo in 1994, as companions to an existing facility in Moline. Deere became conscious of spiraling health-care costs earlier than some other companies, and made moves to become self-insuring as early as 1971. A separate Deere health care division was established in 1978, and the first Deere Family centers opened in the early 1980s. In the new millennium, health coverage of Deere employees was consolidated into a single HMO, and more flexible benefits options were established.

Sophisticated machinery attends to a lot of the work performed on modern assembly lines, but well-trained people still must handle many steps in the process. Unlike past days, when workers toiled in extremely poor conditions, today's line worker benefits from safe and efficient ways of working—a reality that's good not just for employees but for the companies that employ them. At the Deere plant in Davenport, Iowa, seen here, components are raised to levels appropriate to workers' heights—in this instance a tall fellow named Richard Siam. Back strain and other maladies that can result from repetitive tasks are moderated, and absenteeism is reduced.

TOURIST MECCA

CEOs Hanson and Becherer added to their résumés with the development and construction of the John Deere Commons and Pavilion as the centerpiece of the Quad Cities (Moline and Rock Island, Illinois, and Bettendorf and Davenport, Iowa) Riverfront Development project. This was a $50 million civic renewal that broke ground in 1993, and that joined the Deere homestead and company headquarters as an irresistible lure to visitors. The Pavilion itself was opened to the public in 1997, and houses a boggling variety of historic and modern Deere products and memorabilia. The Pavilion has since become one of the top five tourist attractions in Illinois.

In 1962 a team of archaeological students from the University of Illinois discovered the exact spot in Grand Detour, Illinois, where John Deere's blacksmith shop had stood. In this historic place Deere perfected the first successful steel plow that helped set the rest of the company's history in motion. Today, at the John Deere Pavilion in Moline, visitors enjoy a multimedia display of the history of John Deere & Company.

In this painting room at the Deere Collections Center in Moline, vintage tractors undergo one of the final steps of full-on restoration. Old paint is removed from the steel by methods dictated by the level of corrosion, age, and rot. Once scoured of the old paint, the metal is ready for any needed repairs. After priming, the final coats of green or yellow are applied and allowed to dry.

The John Deere Pavilion was conceived as a way to celebrate the history of the company and encourage continued public awareness of the brand. Ground was broken in 1993, and the doors opened in August of 1997. In the first week of operation more than 52,000 people arrived to take in the world's largest display of agricultural manufacturing and history. By 2010, the Pavilion had greeted more than one and a half million visitors from every state in the union, as well as from 50 nations.

Vintage machines and interactive displays have made the John Deere Pavilion a popular Illinois destination. Key exhibits are mounted on a revolving basis, so regular visits are always a good idea. Tractors, a combine viewed in cutaway, and the ways we use food harvested by Deere products are all part of the fun.

CHANGES IN AGRICULTURAL PRACTICES

Deere moved into the 1990s highly conscious of a trend toward "conservation farming." This entailed countering erosion and chemical run-off, regeneration of soil fertility, and steps to address depletion of water tables. For Deere and other agri-equipment manufacturers who looked to the future, the trend encouraged combined, "tandem" in-field operations that meant fewer trips across the soil. Reduced tillage, or no tillage at all, would become common. Traditional fall plowing would probably be eliminated, despite the Biblical admonition, so that roughage and residue would stabilize the soil over winter and prevent wind erosion. Conventional tillage—moldboard plowing, plus up to five additional steps,

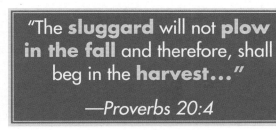

"The **sluggard** will not **plow in the fall** and therefore, shall beg in the **harvest...**"

—*Proverbs 20:4*

such as disking and dragging—can be reduced to perhaps just a single pass with a field cultivator. The main purpose of traditional tillage was weed control, but today that's done mainly through the application of chemicals. But the sad fact is that the cost savings of a no-till approach are offset by the costs of the chemicals and their application.

Today's sophisticated scientific farmer looks to the agri-equipment manufacturer to provide the right tools for the job. Deere's MaxEmerge planter, for example, tilled only a narrow strip for the seed drop with angled coulters that opened a trough, and with wheels behind that compacted the trough closed. Deere sells not just equipment, but *efficiency*.

Once the combines have harvested the crop the stock is passed to a wagon or truck for transport. Once loaded, the trucks move to grain bins or drying silos, where excess moisture is removed prior to sale. The operation seen here took place on the Lee farm in Hardin County, Iowa, in the autumn of 2007.

THE END OF THE FAMILY FARM?

In serious farm country, it has been said that unless you have more than 300 acres, you're just a "hobby farmer." If you have between 300 and 750 acres, you'll need a job in town to supplement your farm income. Upwards of 1000 acres under till are necessary to bring in income sufficient to support both the farmer and the machinery that work the land.

Conglomerate farm companies buy up neighboring farmlands and use hired workers to farm on a grand scale. The small farmer who has inherited the family farm is in competition with these giant businesses. A young person who desires to enter into private-ownership farming is faced with possibly insurmountable financing problems for doubtful financial rewards.

Nevertheless, it is the family farmer who loves his land and is willing to face the incredible challenges posed by weather, market conditions, and government policies. He wants to raise his family and his crops, with the ultimate intention of passing the land on to his children. The family farmer has, for the most part, fed the world, and Deere has been vital to that tradition.

Corn heads affixed to a Deere combine efficiently harvest large fields of corn, and address what farmers call "residue management."

Silage is animal feed created via anaerobic acid fermentation (typically, in a silo) of stalks, chaff, and other byproduct organic matter gathered during and after a harvest. Here, a Deere dumper tips silage into a clamp—a two-walled holding area, often sited on a slope—to feed dairy cows in the winter.

It's critical that "cotton modules" be of uniform size, to facilitate processing and transport. This four-row Deere cotton picker is unloading crop that will be made into bales.

This Deere chisel plow is used in slit tillage, an after-harvest process that reduces erosion of land under heavy corn production. Each curved tooth is about two inches across and spaced 30 inches from its neighbor. The chisel teeth follow a disc coulter between the harvested rows, and dig down 12 to 15 inches.

A Deere combine harvests hardy Claire wheat on Richard Strange's farm in West Sussex, England.

The circular heads of wheeled hay rakes are fitted to adjustable arms that extend from either side of the tractor. Typical work width depends on the model of rake, and ranges from eight to 24 feet. Each wheel is layered with curved metal tines that whip leftover hay from one wheel to the next, creating fluffy piles that will be dealt with later by hay loaders and hay balers.

Deere & Company manufactures hay balers that produce round bales, and others that make square bales. Here, a round baler does its stuff, scooping down on loose hay, processing it through augers, rollers, and teeth that feed it to the forming chamber.

RACING FOR HORSEPOWER, AND OTHER PRIZES

Horsepower is defined as the rate of doing work. As farms have grown in size, as "smart farming" and conservation came to the fore, and as economics continue to demand that as much work as possible be done in as little time as possible, agri-workers know there's no substitute for horsepower. We can only wonder at the patience of the farmers of the 1920s with hundreds of acres to till and only a one-bottom plow.

In 1960, the average full-size farm tractor cranked out 40 to 50 horsepower. By the 1990s, *compact* tractors worked in that range—an achievement that would have been unimaginable 30 years earlier. Large two-wheel-drive tractors of the

'90s had more horsepower than some of the four-wheel-drive types of the '60s. Increases of these sorts might have continued indefinitely except that, by 1990, rising fuel prices forced tractor horsepower to throttle back from peak output figures.

In 1936 Caterpillar's RD-8 crawler was the first tractor to break the 100-horsepower barrier, as recorded by the University of Nebraska. The 34,000-pound tractor had a six-cylinder diesel engine of 1,246 cubic inches. Cletrac soon followed with FD (diesel) and FG (gasoline) crawler models.

The first two-wheel-drive wheel-type tractor to exceed 100 brake horsepower

If Deere has learned anything in its history it's that people buy many sorts of tractors for a boggling array of chores. Here, a John Deere pulls a beach rake to groom sand at a public vacation spot. Similarly, a gang mower is an efficient device to maintain any golf resort's manicured grass. Winter brings a new set of tasks, and Deere rises to the occasion with tractors and other machines designed for basic snow removal or careful grooming of ski slopes.

At the end of the growing season it's the job of the combine to bring in the crops. The wide front end of the machine gathers and pulls the crop into rotating bars. Deere makes combines capable of threshing or separating; the user's choice is determined by the type of crop being brought in. Up to 350 bushels of stock can be stored onboard the combines before the operator extends the unloading auger and passes the material to the auger wagon that rolls alongside.

was the 1962 John Deere 5010. Its 531-cubic-inch six-cylinder diesel made 121 PTO horsepower. To harness this two-wheel-drive's power without excessive slippage, 24.5–32 rear tires were used, with a ballasted weight for the tractor in excess of 17,000 pounds. In 1963 Allis-Chalmers introduced the Model D-21, a two-wheel-drive machine developing 103 horsepower.

Once the 100-horsepower genie was out of the bottle, all the manufacturers were quick to take advantage. The Minneapolis-Moline G-1000 entered in 1965 as a two-wheel-drive model, and was followed in '66 by the 2WD Case 1031. Four-wheel drives came into the picture in 1959 with the John Deere 8010, the International 4300 in 1962, and the Case 1200 in 1964. Deere's 8010, by the way, was the first farm tractor to exceed 200 horsepower.

The Minneapolis-Moline G-706 LPG was the first 100-plus-horsepower tractor tested at the University of Nebraska with front-wheel assist (FWA). Although introduced in 1963, FWA didn't become popular until the 1980s.

Big Bud, a line started in Havre, Montana, in 1968, led the horsepower race for part of the 1970s. Bud's 1976 KT-450 ran with a 450-horsepower Cummins engine.

A one-off Big Bud 1978 model ordered by Rossi Farms of California, the 16V-747, boasted a 16-cylinder engine with a claimed 900 horsepower at the drawbar. And then there was the $1 million Caterpillar D11 crawler, which, at 935 horsepower, remains the industry's horsepower champ. Despite this, the D11 isn't often seen on farms except for deep sub-soiling chores.

At about a million dollars, the 100-ton, 935-horsepower Caterpillar D11 is suited for super-duty applications on corporate farms, but it's most usually seen at mining sites, where rocky surfaces can tear rubber tires to shreds.

Montana-based Big Bud Tractors, Inc., was spun off from a Wagner Tractor dealership owned by Wilbur Hensler and Bud Nelson. Big Bud built heavy-duty machines on custom frames, and came up with some interesting innovations, such as a skid system that facilitated engine removal. This 1989–90 450/50 (*left*) produced a claimed 500 horsepower at the drawbar. Financial trouble forced a takeover by Messner Brothers in 1985. Production of Big Bud tractors was suspended in 1992.

JOHN DEERE
MEMORABILIA

Deere fans who sew enjoy licensed fabrics that can be turned into sheets, tablecloths, and many other items.

This nostalgic miniature in Deere livery imagines an alternate world in which Deere was in the gasoline business.

Enjoy that morning cuppa Joe with a Deere mug.

A nicely rustic, homemade Deere footstool is just the thing after a long day's work.

This handsome mini-locker recalls Deere's centennial with an "1837" number plate, and the 1937 leaping-deer logo. That's a Model B at the bottom.

Here's a sturdy scale model of Deere's mighty 9420 or the similar, slightly larger 9520.

What better way to serve a hearty meal than on official John Deere china?

DRIVE SYSTEMS

By 1960, multi-ratio transmissions were standard; and half-step power down-shifts, like one pioneered by International Harvester in the late 1950s, called the Torque Amplifier, were common. In 1957 Case introduced the Case-O-Matic torque converter transmission, which had excellent load-starting ability. Oliver came out with a similar arrangement in 1958 with its Model 995 GM Lugmatic. Years later, in 1981, Steiger began using the five-speed Allison (then a division of GM) torque converter/powershift transmission with a two-speed auxiliary.

An industry first appeared in 1959: Ford's all-power shift transmission, the 10-speed Select-O-Speed. John Deere offered a similar arrangement in 1963. Today, powershift transmissions with up to 18 speeds are generally available. The Caterpillar Challenger, for example, features a 16-speed forward unit (with nine reverse) providing five shift modes such as pulse shifting one gear at a time, continuous sequential shifting, pre-selected gear shifting, programmable up and down shifting, or automatic shifting through gears 10 through 16.

In 1990 Ford offered a similar 18-speed unit called Ultra-Command. With it, the shift lever was moved forward and backward for forward and reverse in any gear selected.

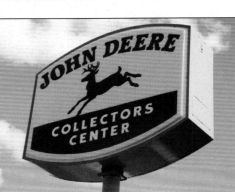

Farms grew larger and larger as the millennium approached, and John Deere stayed current with fresh examples of seriously large tractors. The 9300 was one of them. Produced from 1996 to 2002, it tortured the scales at 31,444 pounds. The articulated chassis was moved via all-wheel drive and a 765-cubic-inch six-cylinder turbo-diesel that developed 360 horsepower. Maximum fuel capacity was 270 gallons, sufficient for hours of nonstop work. The Synchro gearbox gave 12 forward gears and three reverse, while the PowerSync provided 24 forward and six reverse. A Class 3 hitch meant that the 9300 could pull the heaviest implements without taxing itself.

Not surprisingly, the popularity of John Deere implements trickled down to collectors who enjoy scale models of their favorite tractors. The John Deere Collectors Center in Moline, Illinois, housed an inviting array of toys, die-cast models, and Deere promotional items. In July 2008 Deere announced that the Collectors Center would be closed, as part of a consolidation of activity at the popular Deere Pavilion.

In 1969, dairy farmer John Steiger, and sons Douglas and Maurice, incorporated and took his homebuilt-tractor enterprise into a new factory in North Dakota. From the beginning, the company specialized in heavy-duty tractors with sophisticated transmissions, including an automatic that was introduced in 1976. These were novel innovations, but because it was a small company, Steiger—unlike Deere and other giants—hadn't the resources to weather the agricultural downturn of the early 1980s. Bankruptcy and a 1986 sale to Tenneco followed. The machine seen here is a 1977–81 Steiger Cougar III PT-270.

Gears were selected by moving the shift lever left for lower gears, or right for higher speeds.

International Harvester introduced hydrostatic drives in 1969 on their 826 and 1026 models. These tractors also were equipped with manual high-low range shifters, but within either range ground speed was infinitely variable at a constant engine speed. Versatile also offered a hydrostatic drive in 1977 on its Model 150.

Oliver topped the ratio chart in 1972 with 18 forward speeds. This was accomplished with a six-speed conventional transmission and a three-speed power shift auxiliary with under, direct, and overdrive. Allis-Chalmers took away those honors in 1974 with a 20-speed arrangement on its 7030 and 7050 models. Massey Ferguson offered a 24-speed arrangement in 1978. Steiger's Tiger IV had 24 speeds in 1984.

Case-IH had a 24-speed unit in 1985, consisting of four gears and a six-speed powershift. Deere offered a 24-speed unit in 1989. The all-time ratio champ, however, was the 1991 Agco-Allis

Model 8630 with 36 speeds: a 16-speed gearbox and a three-speed powershift auxiliary. The 1987 Massey Ferguson 3050 and 3060 came with 32 speeds. This was accomplished with a four-speed powershift, a gearbox with four synchronized gears, plus a high/low range and a shuttle shift. In 1995, White also boasted 32 speeds, but with an eight-speed manual gearbox and a four-speed power-shift auxiliary.

A sea of Model 4855 awaits shipment from Deere's Waterloo, Iowa, factory.

Rescuing the Past

Company founder John Deere slowed his activities after the Civil War, and found pleasure raising cattle and hogs on a farm near Moline, Illinois, the city he had helped to build. In 1873, eight years after the death of his wife, Demarius, Deere was elected Moline's mayor. Later, he and his second wife, Lucenia (Demarius's younger sister), enjoyed winters in San Francisco or Santa Barbara, but "home" was Red Cliff (shown), a mansion in the Italianate Second Empire style. In an instance of appropriate symbolism, the spacious house stood on a bluff overlooking Moline. John died in 1886, and family members lived in Red Cliff until 1934, when the mansion became a 14-unit apartment house that was eventually abandoned to the elements. An ambitious private renovation began in 1996.

New Fuels and Power Delivery

Since the 1960s the tractor industry has abandoned distillate fuels and introduced liquefied petroleum gas (LPG). June of 1968 saw the University of Nebraska's last LPG tractor test. The industry has also left behind gasoline. International's Model 284 was the last gasoline-powered tractor tested at the University of Nebraska. That test was run in May 1978. Since that time, only diesel tractors have been available on the American market, with the exception of small lawn tractors.

Another possible exception is the "new" N-Series Fords, manufactured by N-Complete of Wilkinson, Indiana. These are remanufactured Ford tractors with parts that meet new-tractor specifications. Every such tractor sold carries a new-tractor warranty.

Turbochargers, intercoolers, aftercoolers, front-wheel assist, dual- and triple-drive wheels, and electronic displays had become commonplace by the year 2000. These things were largely undreamt of in 1960. One of the most profound changes went almost unnoticed, however, and that was the disappearance of the tricycle front end. While the narrow or tricycle front remained in industry catalogs for a time, they were not often sold after 1973. Two things account for the demise of the narrow front: the use of chemicals, which largely eliminated the need for the cultivation of tall crops; and the advent of the front-end loader, which seemed to work so much better with a wide-front configuration.

The Deere 5310K was manufactured by Deere in India, and was based on the 5310 made at Augusta, Georgia, from 1998 to 2000. This rolling billboard for KTB Leasing was snapped somewhere between Krabi and Trang, Thailand.

PRODUCT LINES AND FOREIGN EXPANSION

In the early 1960s, Deere began integrating Mannheim, Germany-based Lanz products into the U.S. tractor line. More and more production of mid-sized tractors of the World Tractor concept (tractors well suited to profitable use in emerging agricultural economies) came from Mannheim and other European plants.

By 1987 Deere saw the need to increase production of its smaller tractors, and built a plant for that purpose in Augusta, Georgia. Also in 1987, Deere added to the 55 Series of Mannheim-built tractors in the 45- to 85-horsepower range. The five new models were rated in ten horsepower steps with the Model 2155 (running with a three-cylinder engine) at 45 horsepower and the Model 2955 at 85.

The 2355 and 2555 used four-cylinder engines, as did the 2755, but that engine was turbocharged. The Model 2955 used

Material handling is a task that requires special equipment, and Deere responded to the need with its Telehandler models. The series bowed in the year 2000, with the 3200 and 3400 propelled by the 4045T engine found in the 6410 tractor. That four-cylinder, turbo motor was rated at 104 horsepower. The 3400 had more reach and material capability than the 3200, but was otherwise similar in design. The Telehandlers were assembled at the Zweibrucken, Germany, plant, replacing previous models built for Deere in Northern Ireland by Matbro.

a six-cylinder. In 1988, the Model 3155 replaced the 3150. Front-wheel assist and a sixteen-speed transmission were standard equipment for this six-cylinder, 95-horsepower tractor.

By the 1990 model year, things were looking up for the farmer and for the agri-business community. Deere & Company introduced 15 new tractor models; the largest, the Model 8960, had a horsepower rating of 337.4. Six new compact tractors, with three- and four-cylinder diesels by Deere's Japanese partner, Yanmar, represented the other end of the scale. One of these, the Model 955, used a hydrostatic transmission. All but one were built by Yanmar in Japan; the 955 was built at Deere's Horicon, Wisconsin, facility.

Deere's long history of innovative product design—and creative marketing efforts over the years—created a collector's market for both antique tractors and promotional items. A few items, like this ad from a short-lived 1997 campaign, may benefit collectors more than they did the company. The "Jenny Deere" campaign, which focused on parts and service, featured an animated, lipstick-accented tractor mascot. But the program was abruptly canceled shortly after distributors received hats and posters. Despite the reversal in advertising strategy, Deere's worldwide annual sales topped $3 billion that year.

The John Deere plant in Mannheim, Germany, was busy in the 1990s. One of its products was the 6400 (*right*), which was motivated by a four-cylinder, turbocharged diesel that produced 85 PTO horsepower. The tractor was sold in two- or four-wheel-drive variations, and listed two different gearboxes: the SynchroPlus, with 12 forward speeds and four reverse, and the PowrQuad, with 16 and 12, respectively. When prepped for action the 6400 weighed just over five tons. Production ran from 1992 to 1998.

A baling machine that labors behind a Deere tractor is a common sight in the fields. The Deere Model 6900 tractor, produced in the Mannheim factory during 1994–97, ran with a 414-cubic-inch turbocharged six developing 130 horsepower, so the British-made Claas Quadrant 1150 baler posed no challenge. The baler's duty is to collect hay from the ground and neatly wind it into a tight bundle. The resulting bales can be easily transported, or can be left in the fields for cattle to feed from.

A Model 4955 underwent testing at the Nebraska Tractor Test Lab in 1989.

Deere bought the Mannheim, Germany-based Heinrich Lanz Tractor Company in 1956. The Deere factory there grew to become the largest exporter of tractors in Europe—it accounts for more than half of Germany's tractor production. Here, a truck driver steers his load of new Deere tractors out of the company's huge Mannheim works.

JOHN DEERE
MEMORABILIA

Nothing turns on the lights like a Deere, either.

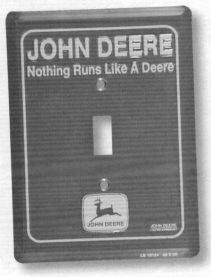

Baby needs a new tractor, c'mon seven!

Deere's 8000T series ("T" for "tracked") came on line in 1997. It was a durable machine designed to handle any light- to medium-duty agricultural chore.

NEW 160- to 225-hp John Deere 8000T Series Tractors

It's not the similarities that upset our competitors . . .

Another alternate-universe artifact: a John Deere locomotive and flatcar, with vintage Deere tractors trailing behind. This quality, HO-scale set by Athearn also included a caboose, a tanker car, and a boxcar.

GREEN
MAGAZINE
June ■ 1994

Deere's mighty Model 630 is featured on the cover of this 1994 issue of *Green Magazine*.

A scaled-down, battery-powered Gator lets kids travel in style.

This hand-some metal and plastic platform-stake truck is modeled after a Chevy from the very early 1950s.

A vintage Deere engine was adapted to power this beautiful ice cream maker found in Tiger, Georgia.

Future farmers don't leave home without their official John Deere jackets.

A MARKET BLIP, AND THAT PESKY COMPETITION

A market downturn in 1991 caused Deere to close its East Moline foundry; this was the first closing of a Deere plant since 1970. The state of the general economy was blamed for the round of consolidations, such as Fiat's purchase of Ford New Holland and the sale of Deutz-Allis to AGCO. Nevertheless, John Deere announced the upgrading of its larger-frame group of row-crop tractors—now called the 60 Series—for 1992, and in 1993, upgraded its mid-sized tractors (those developing 66 to 145 horsepower). The very large articulated tractors were also improved, with the largest, the Model 8970, upgraded via a 400-horse Cummins engine. These 70 Series tractors had electronic engine control that increased the power output of the engine as it lugged down by boosting the intake pressure and fuel flow. This feature was named "Field Cruise."

In 1992, the first off the assembly line of the new Augusta, Georgia, facility was the "Thousand Series." In 1993, the 5000, 6000, and 7000 Series tractors boosted Deere's market share in the U.S. and Europe. Buyers in Germany, for instance, could choose from among tractors made by 20 companies, yet Deere bolted upward from number three to number one in tractor sales in that nation.

Horsepower was one reason. The 5200 had 40 horsepower, the 5300 had 50, and the 5400 ran with 60 hp. In 1996, Carraro of Rovigo, Italy, built some of these models.

Deere's interest in foreign markets grew keener. In 1994 Deere arranged with a well-established Czech Republic company, Zetor, to manufacture a simple, small tractor for developing markets. The arrangement was beneficial to both companies: By 2009, more than a million Zetor tractors were in use in 100 countries around the globe.

For the 1997 tractor lineup, Deere introduced rubber tracks as an option for its 8100 through 8400 Series tractors, which were otherwise four-wheel drive. The power range was 160 to 225 horsepower. Thus, Deere & Company launched its challenge to Caterpillar. By model-year 2000, the Model 8400T had

become the 8410, still at 225 horses, but the Model 9400T, with a horsepower rating of 425, had been added. The 9400T was a companion to the Model 9300 articulated four-wheel drives of the same power output.

International expansion is always expensive, but if it's played properly, the payoff can be tremendous. Deere established the rules of its own game and increased its reach when it looked eastward in 1997, with the purchase of an equity stake in Jialian Harvester Company, a Chinese maker of combines.

As the millennium approached, Deere enjoyed annual overseas sales of more than $3 billion. This was greater than the

Deere's 89-horsepower 5510 was built at Augusta, Georgia.

entirety of the company's sales before the mid-1970s.

Net earnings reached $1 billion for the first time in 1998. That same year, ground was broken for a new tractor-making plant near Pune, India (the facility opened in 2000). Closer to home, Deere acquired Louisiana-based Cameco Industries, establishing a presence in sugarcane harvesters.

Special Technologies Group, Inc. was founded in 1999 as a Deere subsidiary, and was headquartered in Atlanta, Georgia. It grew to offer products and services related to wireless communication, information management, and Internet applications. It also provides electronics and software for vehicle communication and global positioning.

Although the American farm economy experienced a significant downturn in 1999, Deere posted yet another profitable year. This success was a vindication of Deere management and the systems that were in place.

Hans Becherer retired in 2000 and was followed by Robert W. Lane, who brought a broad range of company experience in the global marketplace as well as a previous tenure in the banking industry.

Deere acquired Timberjack, a Finnish manufacturer of heavy forestry equipment, in 2000, and was granted a banking license in Luxembourg, which allowed Deere to finance equipment across Europe. In addition, 2000 saw the establishment of Deere credit offices in Brazil and Argentina.

Without doubt, Deere was now multinational and multivaried.

Built at the Waterloo, Iowa, plant between 1992 and 1996, the Deere Model 7800 was a popular tractor at home and abroad. The 466-inch turbocharged diesel delivered 145 horsepower to the PTO. PowrQuad was the standard transmission, with 16 forward ratios and 12 reverse. The optional PowerShift setup claimed 19 forward ratios and 12 reverse, but as in all PowerShift versions gear changes could be made without using the clutch. The 7800 weighed a scant 9,000 pounds when ready for duty and sold for $84,000 at the end of its production run in 1996. This one pulls a Tyler fertilizer spreader.

From 1994 to 1998 Deere purchased small, low-cost tractors from Zetor, a Czech company, and sold the machines under the Deere name through the company's growing distributor network in Latin America. Gleaming Zetor Silver-type tractors (*foreground*) and Forterra models (*background*) were displayed at this Royal Welsh Agricultural Society Show at Builth Welles, Wales.

In the world of gigantic tractors, the John Deere 8970 of 1993–96 (a '93 is pictured) was one the biggest. Riding a 143-inch wheelbase, this articulated-chassis machine weighed nearly 34,000 pounds when the required 220 gallons of diesel fuel and 72 quarts of coolant were added. An octuplet of massive rubber wheels gave the 8970 plenty of traction regardless of terrain or payload. The 855-cubic-inch turbocharged diesel developed 400 horsepower at the PTO. Deere's Syncro transmission was the standard gearbox, with 12 forward ratios and three reverse. The optional PowrSync setup brought an additional 12 forward gears. In 1993 the 8970's price tag was nearly as big as the tractor itself: $145,000.

Allis-Chalmers, a longtime Deere competitor, wasn't faring well at all in the early '80s and ended up being purchased in 1984 by Deutz of Cologne, Germany. Deutz, in turn, was absorbed into the newly created AGCO in 1990. Before the Deutz-Allis name disappeared for good, it appeared on tractors that included this circa 1989 9130, a versatile, six-ton machine that generated 135 PTO horsepower.

With a trailer full of clippings in tow, this self-propelled 6910 forage harvester from about 1996 is off to the storage facility to complete the silage-cutting process. Even smallish farm operations can generate silage by the ton, which is used not just for livestock forage but for anaerobic digestion, a process by which microorganisms break down the fermenting grass in an oxygen-free environment, creating biogas that can be used for heat and electricity. In addition, the solid, nutrient-rich digestate that remains after anaerobic digestion is a common ingredient of quality fertilizer.

The big 7000 series stormed onto the scene in 1992 with the 7600, 7700, and 7800. The 7800, seen here, was the most powerful of the three, running with a turbocharged, 466-cubic-inch six that was coupled to a PowrQuad transmission. Sixteen forward gears and 12 reverse were on hand, with easy access provided by Deere's Power Shift setup. When equipped with four-wheel drive the 7800 tipped the scales at 15,560 pounds, a capacious weight that included 91 gallons of fuel. The spacious cab could be entered from either side and offered AM/FM radio and other amenities not typically found on farm implements. When the 7800 ceased production in 1996 its MSRP was $84,000.

The Model 6850 is a highly specialized, self-propelled forage harvester. These machines utilize a cutterhead or flywheel to chop the material on the ground into silage before passing it to the onboard storage area. Once passed through the harvester, the silage can be treated with additives that accelerate fermentation. Much like a crop harvester, the 6850 and similar Deere forage harvesters can accommodate hundreds of bushels before having to offload to an alternate wagon or truck.

A Deere 650 disc setup is tugged by a Model 9000 articulated tractor. The 9000T variant runs on tracks.

The 8410T, an 8410 variant, was a dominant tracked machine of the early 2000's. Its 496-cid turbo diesel produced 270 horsepower at the PTO. An electronic control system linked to the engine helped prevent inadvertent operator overload of the drivetrain.

World Leadership

"John Deere is an authentic American folk hero, whose legend began in his own lifetime..."
—*Business historian Wayne Broehl, Jr., 1984*

GLOBAL GROWTH

In 2002, a respected business newspaper, *Crain's Chicago Business,* announced one finding of a nationwide survey: The most trusted Illinois company was John Deere. And in the same year, *Business Ethics* magazine named Deere & Company one of its 100 Best Corporate Citizens. A few years later, in 2007, *Ethisphere* magazine included Deere on its list of the World's 100 Most Ethical Companies. In a period that uncovered awful corporate malfeasance in American banking, energy, and other sectors, Deere's hard-won reputation was golden, indeed.

The firm's reputation for honesty and integrity is particularly noteworthy because of Deere's now-international reach. Major operations are based in Latin America, East Asia, and Australia, and the company maintains its aggressive European manufacturing and sales presence. Today Deere represents not just itself, but the United States, to customers around the globe.

This remarkable growth has been carefully choreographed. In 2001, Deere & Company acquired horticultural supplier McGinnis Farms, of Alpharetta,

> *"...[W]e see our **investment** of more than $2 million a day in research and development as an **essential means to achieve** and sustain strong economic performance, while making **our customers more productive than ever."***
>
> —*Deere & Company chairman and CEO Robert W. Lane, March 2008*

Georgia. Almost simultaneously, Deere purchased New York-based Richton International Corporation, acquiring 170 branches of a Richton subsidiary, Century Rain Aid, a distributor of irrigation and landscape products. Those outlets joined 50 branches formerly owned by McGinnis to create John Deere Landscapes, a new distribution entity with more than 200 branches in 37 states and Canada, serving homeowners and professional landscapers.

The acquisitions of McGinnis and Richton bore even more fruit in '06, when Deere became the USA's leading wholesale distributor of nursery, irrigation, lighting, and landscape materials.

With 255 horses at the PTO, and a weight exceeding 23,000 pounds, the Model 8520 isn't fazed by heavy road-hauling chores. The standard rear-lift capacity at the three-point hitch is 15,180 pounds; an optional setup brings that figure to 17,300. The operator can make use of 16 forward gears and four reverse to get the most from the six-cylinder diesel engine. This one is fitted with optional driving lamps and extended side mirrors. The 8520 was manufactured at Deere's Waterloo plant from 2002 to 2005.

Deere's Model 8230 dates from 2006 to 2009. It's a hardy row-crop machine that runs with a 265-horse diesel producing 232.5 maximum hp at the PTO. A well-equipped 8230 will have front xenon lights, FM business band radio, a 60-gpm (gallons per minute) hydraulic pump, front fenders—and even a dual-beam radar unit to keep precise track of distance and ground speed, area covered, and area per hour. All of that matters because the operation of many modern planters, fertilizer applicators, sprayers, and seeders is controlled by radar speed. In independent analysis, the Nebraska Tractor Test Lab found that the 8230 was nearly 25-percent more fuel efficient than the comparable New Holland TG245 and Case I-H Magnum 254.

Introduced in 2001, the 6020 series was built in Deere's Mannheim, Germany, plant for sale in Europe and the USA. The series included five four-cylinder models, plus five with bigger six-cylinder motors. Of the four-cylinders, the 6420S (*shown*) was at the head the class. A turbocharger helped produce 120 horsepower that fed through a stepless transmission offering seamless gear changes and almost endless ratios. Premium Plus versions featured AutoPowr that allowed a top speed of 31 mph. Two-wheel drive was standard, with four-wheel an available option. The cockpit of the 6420S looked like a gamer's paradise, with a wealth of electronics and control levers.

Deere solidified its standing in this segment with the 2007 acquisition of LESCO, Inc., an important, Cleveland-based supplier of products for lawn care, pest control, landscaping, and golf courses.

A 2003 agreement with The Home Depot placed Deere riding mowers into that retail chain, marking the first time the mowers were available in outlets other than John Deere dealerships.

A late-1980s supply arrangement with the U.S. Department of Defense for articulated tractors for use by the Marines remained active years later, and earned for Deere the DOD's first-ever "highly successful" rating in 2003. Late in 2006, John Deere Construction & Forestry announced a $47 million contract with the Department of Defense for 300 Tractor/Rubber Tired/Articulated Steering/Multi-Purpose (TRAM) units, to be based on Deere's well-liked 624J Wheel Loader.

Meanwhile, international operations continued to grow. A new tractor factory at Montenegro, Rio Grande do Sul, Brazil, was announced in 2004. A year later, Deere opened a seeding-equipment plant in Orenburg, Russia, and assembled a dealer network to service it. Also in 2005, a new John Deere Technology Center near Pune, India, was announced. The facility opened with 35 employees; by 2008, 900 people worked there, expanding the global reach of Deere

A TRAM—a Deere 624J modified for military use—is a heavy-duty bucket loader and forklift. For combat-area applications, the cab is armored.

In 2006, Deere opened a tractor-transmission plant in Tianjin, China, a city of 11 million people located about 60 miles southeast of Beijing. A municipality as well as a city, Tianjin has aggressively developed industrial parks for manufacturing, high tech, and export. Besides transmissions, the Tianjin plant builds small tractors in the 60- to 120-horsepower range for sale across Asia, Africa, and other foreign markets.

Thanks partly to the American surge in home buying as the 20th century drew to a close, Deere was encouraged to enter the home-landscape market, with products designed for commercial landscapers as well as homeowners.

Personnel at Deere's engineering-driven Technology Center in Pune, India, look closely at ways to reduce the time needed for product development. The results have ramifications for Deere operations all over the world.

Deere's Construction Industrial Division produces the 344H loader, whose cousins are the less-powerful 304H and 324H. Bucket capacity for the 344H ranges from 1.7 to 2.0 cubic yards.

design and technology. There is additional Deere activity near Pune, as well: a tractor factory that's been noted for its sterling safety record.

Earnings reached new heights for 2006: a lofty $1.7 billion. In that same year, Deere chairman and CEO Robert W. Lane was named "CEO of the Year" by *Industry Week* magazine. Even as the company savored this recognition, it developed its new wind energy unit, managed by John Deere Credit, and opened a new transmission factory in Tianjin, China. Another new China facility, a tractor plant, was opened at Ningbo in 2007.

In general, Deere's China operations concentrate on small-horsepower tractors intended for sale in China and elsewhere in Asia, in Africa, and in the Commonwealth of Independent States (Russia, Armenia, Moldova, Kazakhstan, and others).

Profits and sales surged for 2008, with more than half of Deere's agricultural-division sales occurring outside the U.S. and Canada, notably in Brazil, Russia, and Western Europe. A new line of 45-to-105-horsepower utility tractors helped spur sales.

Because of this global demand, tractor and harvester expansion was announced for Moline and Waterloo, and for a pair of sites in Brazil, at Montenegro and Horizontina.

Despite the severe worldwide economic downturn that became critical in the autumn of 2008, Deere's global growth continued into 2009, when the company announced a joint venture with XCG Excavator Machinery Company in China. The move was designed to give Deere access to fresh markets in Brazil, Russia, India, and China.

Research, development, and marketing of sophisticated GPS-based technology continued at a brisk pace, allowing agricultural and construction managers to monitor performance, usage hours, and general condition of tractors and other Deere equipment from a desktop computer located hundreds of miles from a given worksite.

New U.S. government emissions mandates—the U.S. EPA Tier 4 requirements for 2008–2015—posed a particular challenge for Deere, which must, among other things, work on its diesel engines to cut sulfur emissions while ensuring that the performance that's expected by Deere customers isn't diminished.

Deere accepted this challenge, while also focusing considerable resources on irrigation and water-management technology, and on wind energy projects in Idaho, Illinois, Missouri, Michigan, Oregon, and Texas.

By the close of 2009, Deere had some 50,000 employees worldwide, nearly half of whom resided outside the U.S.

All of the preceding is to suggest a very large and diversified company, but the core of Deere—or the heart, if you prefer—remains the company's tractor operations for the American market. Deere's preeminence there is enviable, but the Moline-based company is hardly alone on the world's tractor stage.

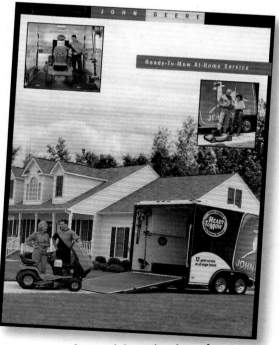

Deere manufactured the Sabre line of "MowMentum" mid-priced lawn tractors and walk-behind mowers.

JOHN DEERE MEMORABILIA

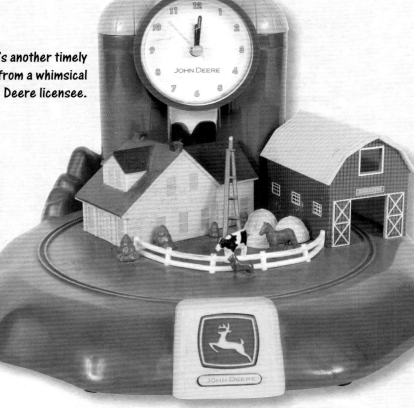

Here's another timely product from a whimsical Deere licensee.

This mouse pad simulating a Deere footplate has a dramatically dimensional look.

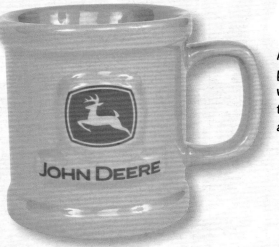

Against all odds, pink goes well with Deere's familiar yellow and green.

Deere has granted licenses to many clockmakers. This piece show-cases the present Deere logo, as well as a modern farm scene.

John Deere came from Vermont to Grand Detour, Illinois, in 1836, and a year later he built this handsome, two-story frame house with a classic portico entrance. Today, the preserved home is shaded by towering elms, and is a popular stop for visitors.

You can't argue with "From small beginnings come great things," the sentiment of this folksy ornament.

Call 'em in for supper with this heavy-gauge, wall-mounted bell.

This traditional blackboard is a pleasing evocation of the past.

THE 21st-CENTURY TRACTOR INDUSTRY

In 2000 four of the world's tractor companies had American names: AGCO, CNH Global (Case-New Holland), Caterpillar, and Deere. Not all of these were American-owned. A great number of the tractors from these manufacturers are made overseas. At its peak in the early 20th century, the industry offered as many as 200 unique tractor brands.

Here's where the industry stands in the first decade of the new millennium:

AGCO

On November 1, 1960, the White Motor Corporation of Cleveland, Ohio, bought out South Bend–based Oliver Corporation. The next year, the Canadian Cockshutt outfit was added to the White family. Minneapolis-Moline was added in 1963. These brand names continued until 1969 when all were dropped in favor of the name White Farm Equipment.

The Allis-Chalmers name was added in 1990 when the German firm KHD sold its Deutz-Allis division to AGCO, which was by then the owner of the White line.

During 1993–94, AGCO purchased the worldwide holdings of Massey Ferguson, which included McConnell Tractors. Today, tractors produced under the AGCO banner are branded AGCO Allis, AGCOSTAR, Massey Ferguson, White, or Landini.

CNH Global, Harvester, and Ford

The November 1999 merger of the Case Corporation and New Holland produced CNH (Case-New Holland), which was at the time the world's largest maker of farm equipment. However, the U.S. Justice Department required some divestitures before approving the merger. Case retained International Harvester but had to sell its ownership in Hay &

Massey-Harris and Ferguson merged in 1952, creating Massey Ferguson. The U.S.-German conglomerate AGCO acquired Massey Ferguson in 1993–94, and still produces tractors with the MF name. This 2009 8600 Series prototype developed 365 horsepower from a 6-cylinder Sisu Diesel engine. Production models were slated for the European market.

AGCO's DT275 runs with a 6-cylinder Sisu Diesel rated at 320 engine horsepower, and 275 at the PTO. The tractor is manufactured at Beauvais, France, and is one of many products from AGCO and others that motivates Deere to aggressively pursue markets across Europe.

International Harvester's agricultural operations were merged with Case in 1985. Today, I-H farm tractors, like this 2007 Maxxum 140, are branded "Case IH." The 140 has a 140-horsepower engine, and ran with 17 forward speeds and 16 reverse. All-wheel drive was optional.

England, in order to expand its presence in Europe.

In 1985, the Case parent company, Tenneco, acquired International Harvester's Tractor and Implement Division, and folded it into Case. The consolidation put Case-IH on a firm foundation for the future and allowed the purchase of the ailing Steiger Tractor Company of Fargo, North Dakota.

International Harvester had been number one in tractor production until 1963, when John Deere took the lead. Harvester countered with thoroughly modern tractor designs, but I-H management elected to keep stockholders happy over the short term with generous

Forage Industries, which was jointly held with AGCO Corporation. New Holland, which owned Ford Tractor Division, as well as New Holland Farm Equipment Company and Versatile Farm Equipment Company, sold off its Winnipeg, Canada, plant along with its Versatile and Genesis tractor lines. New Holland was allowed to keep its Versatile TV-140 bi-directional tractor, which could be operated from the cab end or from the engine end.

Although CNH is headquartered in Racine, Wisconsin, Italian auto-making conglomerate Fiat holds a 71-percent ownership stake.

The entities that now comprise CNH started on the road to takeover during the political and economic upheavals of the 1960s and '70s. Oil and grain embar-

goes, and interest rates of 15 percent and higher, caused farm prices to plummet. That, in turn, diminished the ability of farmers and agri-businesses to purchase new equipment.

J. I. Case rode out the 1960s fairly well, mainly on the strength of its 1957 acquisition of the American Tractor Corporation, which put Case solidly into the crawler tractor and construction equipment businesses. But Case had a history of management that was alternately deeply conservative and sharply flamboyant. By 1970, cash shortages forced Case to sell out to a giant conglomerate, Tenneco, Inc. The sale worked out well for Case, which was then able to issue a bevy of technologically advanced tractors. In 1972, Tenneco acquired David Brown Tractors of

New Holland dates back to the 1890s, when the company operated a grain mill and made farm implements. It was purchased by Ford in 1985, and became part of Fiat five years later. The New Holland name carries on today, as part of CNH Global, which also includes Case and International Harvester. This is a 2009 New Holland TD5050 high-clearance tractor, designed for vegetable and specialty-crop growers.

XCG Excavator Machinery Co. is a Deere partner based in Jiangsu, China. It specializes in the import and export of heavy machinery, like the 23-ton Model 2301 excavator seen here. This machine makes use of components from a variety of manufacturers that includes Cummins (main engine), Kawasaki (main pump, swing motor), and Hy-Dash (drive motor). XCG allows Deere access to numerous foreign markets.

dividends, rather than to plow enough profits back into development of plants and new technology. Even at that, stockholder unrest caused several management shakeups as dividends fell. In 1979, the United Auto Workers Union struck I-H for higher wages and increased benefits. The strike lasted nearly six months, and Harvester didn't recover until its sale to Tenneco six years later.

After 80 years of tractors from the Ford Motor Company, the name disappeared from those machines in 1997. Ford had acquired New Holland in 1986 and the Ford Tractor Division

became Ford-New Holland. In 1994, the Fiatagri Group began a buyout of Ford-New Holland, which was completed in 1995. Since 1997 the line of tractors has been marketed worldwide under the New Holland brand.

For years, Ford was a two-tractor company, riding high throughout the early postwar marketplace with just two tractors, the Fordson and the 8N. In their heydays, the Fordson and the 8N each were capable of selling more than 100,000 units in a single year. These were far and away the best-selling tractors in America, until the market and Ford began to change. Ford Motor Company went public in 1955 and Henry Ford II labored to satisfy stock-holders. A surprise came in 1956, when British Ford Tractors outsold the domestic unit. Rather than be further embarrassed, Ford's American tractor arm expanded the domestic line to two engine sizes with a variety of transmission and other options, including diesels and LPG (liquefied petroleum gas) fuels. U.S. sales still slumped, so the concept of the "World Tractor" was developed in 1961. By 1964, a new line of Ford tractors was being made in new plants at Highland Park, Michigan, Basildon, England, and Antwerp, Belgium.

The World Tractor concept was successful and by 1966 Ford was again number two in tractor sales. In 1977, to celebrate 60 years in the tractor business, Ford marketed articulated, four-wheel-drive monsters built by Steiger. Ford continued to expand with the acquisition of Sperry's New Holland Group and the Canadian Versatile Company just before agreeing to the sellout to Fiat.

Caterpillar

Caterpillar, like Deere & Company, has sailed through the last 30 years virtually unscathed. Both companies have long histories of exemplary management and outstanding products. After a period of close market cooperation, Deere decided to use its Lindeman conversions to compete with Caterpillar crawlers. Soon after, Deere's Industrial Division went after a piece of Cat's heavy-equipment market.

Although Caterpillar had generally focused on heavy equipment, it made a bid for increased farm sales in 1987, when the rubber-tracked Challenger farm tractors arrived on the scene. These machines combined the advantages of rubber-tired and track-type tractors with good highway transport speeds and high flotation over soft ground. Deere has responded to the challenge in recent years, with rubber-track tractors of its own. CNH and other manufacturers have done likewise.

Although Illinois-based Caterpillar made its fortune with heavy equipment, it launched a big push for increased farm sales in the late 1980s. In 2002, however, Cat sold its agricultural division to AGCO; today, AGCO Challenger tractors are sold through Caterpillar dealerships. The D6N Waste Handler seen here is intended mainly for industry, but may be practical on very large farms that need machines for landfill duty. The D6N is protected with engine shields and an uprated cooling system.

Deere Organization and Management

Today, the company founded by John Deere is broadly divided into two businesses: Equipment Operations and Financial Services. Deere's 2008 annual report shows a gross income from both businesses in excess of $28 billion. Equipment Operations contributes to that income by manufacturing and distributing a full line of agricultural equipment; a variety of commercial, consumer, landscape, and irrigation products; and a broad range of equipment for construction and forestry.

Financial Services primarily provides credit to finance sales and leases to John Deere dealers. In addition, the financial side offers crop risk-mitigation products and investments in wind-energy technology.

A 2008 issue of *John Deere Journal*, the Deere employee publication, celebrated the bicentennial of John Deere's birthday. *JDJ* is available to the general public on the Deere Web site. The August 2009 issue featured, among other things, a conversation with then-new CEO Samuel Allen, and coverage of the company's efforts to expand global trade.

Don VanSkiver services a Deere engine at Charles Oliver & Son in Canaseraga, upstate New York. Oliver sold Deere products from 1945 to 2008.

The emergence of a U.S. market for renewable-energy sources in the early 21st century led Deere to form John Deere Renewables. The wind-power division of that company sells turbines to agricultural customers in areas where the wind blows frequently enough to make wind power economically feasible—notably in states such as Michigan, which will require, by law, that renewables comprise 10 percent of electricity used in the state by 2015. Deere activated its first Michigan wind farm early in 2008. Overall, John Deere Wind Energy's wind portfolio is capable of producing enough power to service more than 40,000 homes.

John Deere Classic

Golf phenom Michelle Wie and her caddy, Greg Johnston, line up a shot at the 2006 John Deere Classic, hosted by the Tournament Players Club at Deere Run in Silvis, Illinois. Established in 1971, the tournament's purse now exceeds $4 million. Past winners include Vijay Singh, Steve Stricker, and Payne Stewart. Although the Classic is hosted by the PGA, an exemption was made for Michelle Wie in 2005, when she played in qualifying rounds with male golfers. She finished 1 under par, and missed the cut by just two strokes. At Silvis a year later she was 6 over par after the first round, and continued to struggle into the second day. After falling victim to heat exhaustion, she withdrew from play.

JOHN DEERE
MEMORABILIA

For farm-style fun, just open a John Deere party pack. This is a brightly appealing product that even has games and quizzes on the reverse of the placemats.

Cleveland-based American Greetings has licenses for Deere-inspired napkins, note cards, and many other items, including this appealing birthday card for Dad.

For tractor, truck, or car, this ladies' license plate proclaims loyalty to Deere.

A little bit German Gothic, a little bit Old English—that's the nature of the numerals on this starkly dramatic clock face.

JOHN DEERE

JOHN DEERE
LICENSED PRODUCT

NOTHING RUNS LIKE A DEERE®

This pencil is bigger than it looks. Much bigger.

Reproductions of vintage ads for Deere's Model H—as well as a wrench in lieu of a traditional handle—give this lunch box a dynamic look.

Changes the Farming Picture

JOHN DEERE
SULKY PLOW
MOLINE, ILL.

BEAM DRAFT ... HIGH CLEARANCE ... SPRING LIFT ...

JOHN DEERE

Look carefully for the jeans pockets in the blue background of this wall hanging.

DEERE AGRICULTURAL EQUIPMENT TODAY

While John Deere is the producer of a wide variety of products, it remains best known for tractors and other items of agricultural equipment, and is the world's largest manufacturer of such products. The breadth of this Deere segment is mind-boggling.

Deere's agricultural tractor activity is made up of six series, as follows:

1. Compact Utility
2000 Series.............. 24.1–31.4 horsepower
3000 Series.................................. 27–44 hp
4000 Series.............................. 40.5–66 hp

2. Utility
5D/E Series .... 45–75 hp
5E Limited.............................. 83–101 hp
5M Series.................................. 65–105 hp
6030 Series................................. 75–95 hp
6D Series.............................. 100–140 hp

3. Row Crop
7030 Small Frame................... 100–140 hp
7030 Large Frame.................. 140–180 hp
8R/8RT Series .... 225–345 hp

4. Four Wheel-Drive Articulated
9030 .... 325–530 hp
9030 Scraper Special .... 425–530 hp

5. Track-Type
8RT Series .... 295–345 hp
9030T Series .... 425–530 hp

6. Specialty Tractors
(Orchard, Greenhouse, Nursery, and Vineyard)
A and F Series............................ 21–96 hp
5095 Hi-Crop .... 95 hp
5105 Low Profile .... 105 hp
5EN Narrow Series................... 83–101 hp

An impressive variety of Deere products awaits purchase at Larson Farm & Lawn in Rogersville, Missouri.

The Model 5403 is an emerging-markets product manufactured by Deere at its facility in Pune, India. This 74-horsepower machine entered production in 2007.

Other segments of the Agricultural Business side include:

- Greenstar, a satellite-based, computerized piece of equipment that combines navigation function with seed and fertilizer application-rate information, plus many more high-tech solutions to farm-management problems.

- Combines, cotton harvesters, sugar beet harvesters, and hay and forage harvesters.

- Nutrient applicators and sprayers.

- Tillage equipment, planters, and seeders.

- Material-handling equipment and scrapers.

- Cutters and shredders.

- Gator utility vehicles.

- Diesel engines.

- Home workshop hand tools for farm and home.

- Frontier Equipment—small tractors and attachments, from landscaping to loader work; from tillage to snow removal.

"Big Time," for sure: The 9030 Series that bowed in 2007 offered wheeled and tracked (*shown above*) versions that produced as much as 583 horsepower.

Hartland Equipment is comprised of four stores in Kentucky. The franchise maintains a sophisticated Web site, and places considerable emphasis on service. Sprayers (*right*) are among the products offered by Hartland.

Above: Lawn and garden products by Deere await buyers at Platte Valley Equipment in Wahoo, Nebraska.

NEW LEADERSHIP

Chairman and chief executive officer Robert W. Lane retired in June 2009 and was succeeded by Samuel R. Allen, who became president and CEO. Allen is also a member of the Deere Chief Operating Officer & Company Board of Directors.

A 34-year veteran of John Deere, Allen was given the big chair after serving as president of Deere's Worldwide Forestry and Construction Division. Earlier in his career with Deere, Allen worked in positions of

The 300D articulated dump truck (ADT) combines a healthy load capacity of 32.2 cubic yards (maximum) with adherence to federal emissions standards.

Robert W. Lane (*far right*) was chairman and CEO of Deere & Company from 2000 to 2009. His global outlook and a background in finance served him well, as he aggressively expanded the company's international presence. The photo seen here was taken in 2006, when Lane gave an award to golfer Deane Beman, who won the first two Deere Classic events, in 1971 and '72.

With agriculture becoming an increasingly multinational business in the new millennium, Samuel R. Allen, who became Deere president and CEO in June 2009, explored alternative-energy technologies, and ways to increase world food production.

increasing responsibility in domestic and overseas operations.

That was no accident, for Deere & Company is unique among Fortune 500 companies in its manner of grooming executives. A typical corporation may set its eye on promising candidates and place them in jobs that relate directly to eventual leadership of a particular division or of the entire company. Such executives are encouraged to work toward the assigned goal and that goal only. During their careers, the executives may develop a sort of tunnel vision that isolates them from the broader aspects of their company's operations.

Deere, in contrast, is known for shuffling its senior executives among a variety of company divisions, so that they have useful experience with numerous aspects of Deere's activities—hence Samuel Allen's prior experience in Deere credit and finance, human resources and industrial relations, construction, forestry, and power systems. He assumed the CEO job after having direct responsibility for Deere's "Region I" operations comprising the Far East, Australia, Latin America, and South Africa.

The 5320 was built from 2000 to 2004, and followed closely on the heels of the 5310. Both were built at Deere's Augusta, Georgia, facility. Power was provided by a three-cylinder diesel displacing 179 cubic inches. Output of 64 horsepower was coupled to nine forward gears and three reverse. The standard model was two-wheel drive with four-wheel an available option.

MODERN DEERE TRACTORS, IN DETAIL

In 1960, the last year of Deere's two-cylinder line, the company offered six distinct models of farm tractors. By 2009 there were 52 distinct models. In 1960, power ranged from 27 to 215 horsepower; by 2009 the range was 21 to 530 horsepower.

In 1960, Deere Transmissions offered four or five speeds forward and one in reverse (some crawlers had shuttle shifts where reverse was available in all five forward ratios). Nearly 50 years later, setups with eight forward speeds and six reverse (or with infinitely variable hydrostatics) were just the beginning for the smaller tractors; minimum for the larger tractors was 24 speeds forward with six reverse.

Deere's largest tractors came with rubber tires or rubber tracks. There were four-wheel drives with and without articulation, and provisions for as many as three tires on each side of each driving axle. All engines were diesel and ranged from 61 cubic inches of displacement to 765 cid. Three-, four-, and six-cylinder engines were available. Turbos and intercoolers were not uncommon.

Early in the 21st century, this level of model and options variety wasn't unusual in the industry. Case-International fielded 29 models, Ford-New Holland had 42, and AGCO had 52, which by then included Massey Ferguson and several other brands.

Deere is a leader, but innovates and does battle every day to stay ahead.

The 624J diesel loader has a bucket capacity (heaped) of 3.5 cubic yards. The bucket is 105.9 inches across, and within its full range of motion—full-height extension to ground level—it has a load capacity of 17,290 pounds to 29,200 pounds. The U.S. military runs a variant called the TRAM.

Deere's massive Waterloo Tractor Works complex in Waterloo, Iowa, covers more than 125 acres and includes not only tractor production lines but an iron foundry, as well. The production line received a robotic overhaul in the early 1980s, when Deere's workforce in the area peaked at 16,000. But a recession and a series of labor disputes trimmed the number of employees to just over 5,000 by 2008. Like many Deere facilities, the Waterloo tractor line shuts down for one or more weeks during periods of low demand, often in the summer.

JOHN DEERE
MEMORABILIA

JOHN DEERE COUNTRY
NOTHING RUNS LIKE A DEERE.

Two leaping-deer trademarks—present and vintage—are worked into the design of this photo tin.

JOHN DEERE
PHOTO BOX TIN

Equally at home on the barn, in the garage, or on the wall outside a bedroom, this lively tin sign makes the owner's allegiance very clear.

You'll spare the finish of your furniture with these leaping-deer coasters.

A promotional Deere postcard issued by the company's operations in India.

Can a clock shaped like a tractor tire be elegant and understated? This one can.

Ray Mack of Lincoln Township, Michigan, owns dozens of restored Deeres.

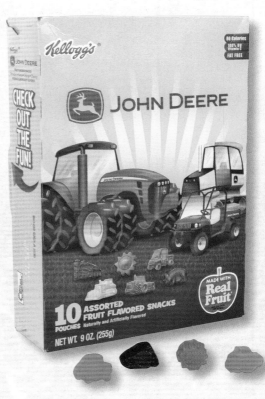

Like Deere, Kellogg's is emblematic of America. The Michigan-based company whips up these colorful fruit chews in the shapes of Deere tractors, 'dozers, pigs, and even the sun.

Mattel's John Deere pink label Barbie arrived in time for Christmas 2007, complete with farm-girl attire, a doll stand, diorama box art, and the coveted certificate of authenticity.

For his personal use, engineer and designer Balaji Rengarajan executed these slick renderings of imaginary motorcycles with the Deere name and logo. Clockwise, from top left, the drawings propose a lightweight street runner, a dirt bike, and a heavyweight touring model.

FORGING THE FUTURE BY RECALLING THE PAST

Integrity, quality, commitment, and innovation—these were the core values of John Deere's company in 1837, and remain the focus of Deere philosophy today. These are the values that have characterized the company in good times and hard times—and there have been plenty of both during the last 170-plus years. Only a few American businesses today can trace their histories to before the Civil War, and fewer still to a time when Indian Wars still raged.

The story of John Deere's company is also the story of America and Americana, and it has played out in the context of nearly two centuries of social, economic, and technological change.

A styled Model A heads a Bob Wills parade in Turkey, Texas, around 2002.

From the westward migration, the transition of territories into states of the Union, to the expansion of the railroad and highway systems; to the great "trusts" of the turn of the 20th century; from wars, panics, depressions, droughts, and political chicanery; to embargoes, the oil crisis, and Japanese and other competition, John Deere has kept its gaze steady. The company patiently weathered each storm as it came, and emerged from the other side the stronger for it.

John Deere may have been tardy in fielding its first farm tractor, but once the Waterloo Boy and the mighty Model D entered the scene, the company never looked back. From then on, the quality and reliability of John Deere tractors burnished the company's reputation and helped it sell an amazing, never-ending variety of ancillary farm equipment—including steel plows, the product that started it all in Moline back in 1837.

Deere's commitment to stand behind its products, and to help the farmer—a small businessman, after all—finance his purchases, has tremendously enhanced Deere's reputation with its primary customers. In particular, the company's willingness to carry the farmer during bad times, rather than dun him for immediate payment, has cemented a bond between the farmer and the Deere dealer. This loyalty extends to the parent company, with a passion bordering on religious fervor.

Finally, Deere & Company is unique in another way. There is at Moline and other locations what is known as the

Startup funds for what morphed into Deere's highly sophisticated "Walking Machine" prototype came from military interests in the Finnish government in the late 1980s, when a Finnish company, Plustech Oy, was tapped to develop a prototype. By 1998, work was based at the Tampere, Finland, headquarters of Timberjack, a Deere subsidiary that controlled Plustech. When Deere acknowledged its connection to the six-legged machine in 2002, the Walker was described as a delicately stepping harvester that could fell timber with minimal trauma to the forest floor (particularly roots and young trees). The articulated, computer-controlled hydraulic legs worked independently of each other, with appropriate placement of each rubber "foot" directed by sensors. The operator controlled the amount of pressure any of the feet brought to bear on the earth, as well as the height of each step. As terrain dictated, overall ground clearance could be adjusted, too. Besides forward and backward movement, the Walker could shift itself diagonally and from side to side. Though remarkable, the Walker ultimately stalled at the prototype stage.

"Deere attitude." It's an attitude that evokes nostalgia. From the executives' mahogany row to the production floors, there is a universal respect for the past. Deere men and women believe in their company, not so much because it is their company, but because they believe the company deserves it. This respect for the past is reflected in employees' attitudes about the future, where Deere's traditional qualities of product and service will be perpetuated. Any shortcoming on the job is regarded as a breach of trust, and a failure to live up to Deere tradition.

There used to be a saying, "As General Motors goes, so goes America." Now, however, it seems appropriate to substitute the name of Deere & Company.

John Deere on the Web

The attractive and functional John Deere Web site, www.deere.com, allows the curious to read about Deere history and important executives, and enjoy pictures and information about new tractors and other equipment. For shareholders and professionals, the site gives quick, practical reference for products and services related to agriculture, construction, and the home, as well as wind energy, crop insurance, credit, parts, and investor relations. The latest Deere press releases and executive speeches are posted on the site, and travelers can get a feel for John Deere attractions. Specialty pages have information about golf and forestry activities, and updated news about federal and military sales. In sum, deere.com is one of the best places to keep current with what's happening at Deere & Company.

LICENSE TO TILL: HOW DEERE BUILT A BRAND EMPIRE

Licensing has long provided Deere with a way to put the company name on products used on and off the farm, and earn royalties for Deere, as well.

Early Deere leaders knew they needed a brand identity for their rapidly growing product line. The "leaping deer" trademark appeared in the 1870s, partly to prevent imitators from diluting the success of the "Moline" plow and other products being sold through Deere branches in the Midwest. Branch managers, who sold other companies' products, too, learned which items farmers wanted most. In time, Deere struck licensing deals to sell those more popular products, sometimes as a prelude to outright purchase of the companies themselves.

Some Deere licensing efforts were structural in nature. In 1879, Charles Deere and Alvah Mansur formed Deere & Mansur to develop new products without risking John Deere's family finances. The new company added Deere-labeled hay rakes, stalk-cutters, and other implements to the product line. Linking the Deere name—with an emphasis on product quality—to new product lines became a recurring theme in company history.

Other early products made by others but bearing the Deere name included the Hawkeye Riding (Sulky) Cultivator, patented by Robert W. Furnas, and bicycles sold during a cycling craze at the end of the 19th century.

In modern times the Deere name has seemed to be almost everywhere: on Deere work boots, made by Dan Post Boot Company; John Deere insulated drinkware by Tervis; soybean-based foam cushions in Ford and Deere vehicles; as well as mowers, chain saws, and financial services. Deere also used a combination of acquisitions and licensing deals to enter European, Asian, and Latin American markets.

Through marketing giveaways and licensed items not directly related to farming, Deere has not only reinforced its brand but also created a market for an astonishing array of products and collectibles, including belt buckles, toys, key chains, playing cards, pocket knives, Christmas ornaments, pajamas—even a *John Deere: Harvest in the Heartland* video game for Nintendo fans.

INDEX

PICTURE CREDITS

FRONT COVER: **Charles Freitag**
BACK COVER: **Ed Bock/Corbis**

AGCO Corporation: 174 (right); **Alamy Images:** Malcolm Case-Green, 151 (top); Nigel Cattlin, 150 (bottom center); Clynt Gamham Agriculture, 168 (bottom); Robert Convery, 163 (bottom); David R. Frazier/Photolibrary, Inc, 147 (left); Grant Heilman Photography, 150 (bottom right); Doug Houghton, 151 (bottom right); Kim Karpeles, 148, 156 (right); Philip Lewis, 150 (top); RIA Novosti, 77 (top); Jim Ringland, 146 (bottom left); George Robinson, 151 (bottom left); Tim Scrivener, 150 (left); Doug Steley, 21 (left); Stephen Saks Photography, 186 (bottom); T.M.O. Pictures, 161 (top right & bottom right); Peter Titmuss, 152 (top); Jim West, 177 (bottom); **AP Images:** 141, 177 (center), 183 (bottom), 185 (top left); **Art Resource:** New York Public Library, 26 (bottom); SEF, 127; **Steve Ballard:** 34; **Joerg Boethling:** Agenda, 170 (bottom right); Peter Arnold Inc., 159 (bottom right); **David Bordner Collection:** 47 (center & right center); Corbis: 94 (right); Bettmann, endsheets, 18 (top), 23 (bottom left), 28 (right), 30 (right), 31 (right), 32 (left), 33 (right), 75 (top), 97 (top left); Rick Dalton/AgStock Images, 143; Gehl Company, 77 (bottom right); Lake County Museum, 97 (bottom center); Rick Miller/AgStock Images, 8; PEMCO - Webster & Stevens Collection, Museum of History and Industry, Seattle, 73 (right); Louie Psihoyos/Science Faction, 157 (top right); Ed Young/AgStock Images, 166 **Curt Teich Postcard Archive:** 97 (top right, top center & bottom right); **Defense Industry Daily:** 170 (top left); **Fernando del Real:** (top right); **David Drew:** 153 (top); **Robert Elzey:** 161 (bottom left); **FarmPhoto.com:** 146 (top right), 153 (bottom), 157 (top left); **Planeta Gadget:** 186 (top); **Getty Images:** 18 (bottom), 23 (right center & bottom right), 24, 32 (right), 42, 147 (top left), 177 (left), 182 (left center & bottom); Time Life Pictures, 16 (bottom), 147 (bottom); **Greg Goebel:** 182 (top); **The Granger Collection:** contents, 16 (top), 19, 22, 23 (top), 28 (left), 31 (center), 35, 94 (left); **Courtesy HA.com Photography:** 38 (top), 131 (top right); **Hartland Equipment Corporation:** 181 (left); **High Contrast:** 175 (top); **Mark Igleski:** 126 (bottom right); **Institute for Regional Studies and NDSU University Archives:** 44; **The Image Works:** 55; **IMRE Communications:** 181 (right); **John Deere Web site:** 187 (bottom); **KS Construction:** 176 (top); **Larson Farm & Lawn, Inc.:** 180 (top); **Library of Congress:** contents, 33 (left); **M. Hanna Construction Co., Inc.:** 183 (top); **MachineFinder.com:** contents, 180 (bottom), 181 (top); **McLean County Historical Society:** 85 (bottom left); **Glen A. Martin/Martin's BikeShop, Inc.:** 56 (bottom left & bottom center); **Minnesota Historical Society:** 41 (bottom), 45 (bottom left), 75 (bottom right), 118 (bottom); Duane Lundquist, 129 (top left); Gordon Ray, 62, 108; **Missouri Valley Special Collections, Kansas City Public Library:** 41 (top); **Andrew Morland:** contents, 3, 9, 17, 39, 43, 49 (top), 50, 58, 60, 61, 64, 65 (top & bottom left), 66, 67, 68 (top), 69 (top & right center), 72, 73 (top), 76, 82, 83, 87, 89 (left), 90 (bottom), 91 (top left, top right & bottom left), 95 (top), 96, 99, 100, 106, 107 (top left), 109, 111, 113, 116, 117, 119, 122, 123 (bottom left & bottom right), 124, 125 (bottom left & bottom right), 128 (top & right center), 138, 139 (top & bottom), 156 (left), 158 (right), 159 (top & right center), 162, 163 (top), 164, 165 (top right, bottom left & bottom right), 168 (top), 169, 171 (top), 182 (right center); **National Archives, and Records Administration:** 134; **Nebraska Tractor Test Laboratory, Records, Archives & Special Collections, University of Nebraska-Lincoln Libraries:** 128 (left center), 132 (bottom right), 159 (bottom center); **Photofest:** Buena Vista, 147 (top right); **Picasa:** 158 (left); **PIL Collection:** contents, 10-11, 20 (top), 25, 26 (top), 29, 37, 38 (bottom), 47 (top left), 48 (left & bottom), 53 (top left & bottom left), 56 (top left, top center, top right & bottom right), 57 (top left, top right & bottom right), 59, 63, 65 (bottom right), 68 (bottom left), 69 (bottom center & bottom right), 70, 71, 74, 78, 79 (top right, center, bottom left & bottom right), 80 (bottom), 81, 84 (top left & bottom), 85 (top & right), 86, 88, 89 (right), 90 (top), 91 (bottom right), 92, 93, 95 (bottom), 98, 101, 102, 103, 104, 105 (top left, top right & bottom right), 107 (top right & bottom), 112, 114, 115, 118 (top & right), 120, 121, 123 (top), 129 (top right, center, bottom left & bottom right), 130, 131 (top left, bottom left & bottom right), 132 (top), 133, 135, 136, 137, 139 (right), 140, 142, 144, 145, 146 (center & bottom right), 154, 155, 159 (left), 160, 161 (top left & right center), 165 (top left), 170 (bottom left), 171 (bottom), 172, 173, 177 (top right), 178, 179, 181 (center & bottom center), 184, 185 (top right & bottom left), 187 (top); **Platte Valley Equipment:** 181 (bottom right); **Robert N. Pripps Collection:** 12, 13, 14, 15, 20 (bottom), 21 (right), 27, 31 (left), 36, 46, 48 (top right), 49 (bottom), 51, 52, 53 (right), 73 (bottom left), 75 (bottom left), 77 (bottom left), 80 (top), 84 (top right), 110; **Andrew Raimist:** 126 (bottom left); **Balaji Rengarajan:** 185 (bottom right); **Rock Island County Historical Society:** 30 (left), 45 (top & bottom right), 47 (bottom right), 57 (bottom left), 132 (bottom left), 157 (bottom); **Courtesy Orion Samuelson:** 6, 7; **SuperStock:** age fotostock, 149, 152 (bottom); **Thomasnet News:** 176 (bottom); **Tractor Pool:** 175 (bottom); **Van's Implement, LTD:** 174 (left); **Wisconsin Historical Society:** 54, 68 (right), 69 (left), 79 (top left), 105 (bottom left), 146 (top left); **Yale University Manuscript and Archives:** 125 (top), 126 (top left & top right)

Additional Photography: Vince Manocchi; Doug Mitchel